# PHYSICS PROBLEM-SOLVING TECHNIQUES FOR UNDERSTANDING AND SUCCESS IN FIRST YEAR MECHANICS

# PHYSICS PROBLEM-SOLVING TECHNIQUES FOR UNDERSTANDING AND SUCCESS IN FIRST YEAR MECHANICS

## A STRUCTURED APPROACH FOR SCIENTISTS AND ENGINEERS

FIRST EDITION

**Christine Berven, Ph.D.**
University of Idaho

cognella®
SAN DIEGO

Bassim Hamadeh, CEO and Publisher
Kristina Stolte, Senior Field Acquisitions Editor
Carrie Baarns, Manager, Revisions and Author Care
Alisa Munoz, Project Editor
Susana Christie, Senior Developmental Editor
Casey Hands, Production Editor
Emely Villavicencio, Senior Graphic Designer
Greg Isales, Licensing Coordinator
Natalie Piccotti, Director of Marketing
Kassie Graves, Senior Vice President, Editorial
Jamie Giganti, Director of Academic Publishing

cognella® | ACADEMIC PUBLISHING

3970 Sorrento Valley Blvd., Ste. 500, San Diego, CA 92121

# Brief Contents

# Detailed Contents

# Acknowledgments

I would like to thank my first physics professor, Professor James Pengra, Ph.D. of Whitman College who ignited my love of physics and showed me how seemingly different problems in physics can be understood using only a handful of basic principles. Without him this book would never have come into existence. I would also like to acknowledge my students who over the years challenged and taught me how to be a better teacher. Sincere thanks are also due to my graduate student Dillon Morehouse for extremely useful editing of the first draft of this book. Finally, thanks also go to the editors at Cognella and the reviewers of the preliminary edition of this book who's honest, and sometimes blunt, comments helped me create the much-improved version you see here.

# Introduction

Two commonly heard comments from struggling physics students are "I understand the physics, but I just can't do the problems," and, "If I only knew what equation to start with, I could solve the problem."

This book was written to help students learn to use a few key principles that will allow them to become expert problem-solvers. They will learn that it is in the initial set-up and analysis of a problem where they combine their logical and basic physics tools to create what is commonly seen as the starting equation. In the process of mastering this skill, they will learn that once they learn how to apply a particular physics principle, they can solve any problem that uses the same principle. It is the initial "setup" process that is the focus of this book, with a central theme that "*a well-understood problem is a problem half-solved.*"

This book is designed and intended to address the shortcomings of first-year physics textbooks and problem-solving supplements that focus too much on the mathematics of problem-solving. A lack of consistent explanations for the process or the reasons for particular steps leaves students trying to infer these on their own. Students then resort to equation hunting in the hope that the one they find will provide an answer, although seldom with any insight gained. In contrast, this book was written to show how to identify the type of problem to be solved, apply simple logic and the mathematical form of the key principle at the core part of the problem, and then solve the problem.[1]

## Audience

The need for this book became apparent while teaching a first-year calculus-based mechanics physics course at the University of Idaho with an enrollment between 120 and 140. Through no fault of their own, quite a few of the students had not learned general problem-solving techniques. Instead, they had only learned how to calculate answers to simple problems. Year after year, anonymous "get to know the class" surveys revealed that even though many had taken and even expressed comfort with Calculus III or even differential equations, word problems were still scary to them.

As such, this book was written with similar students in mind: those taking first-year calculus-based mechanics at the college level who would benefit from learning how to approach problem-solving in a systematic way. Although that was the target audience, this book could easily be of help to students in high-school physics and college algebra-based first-year mechanics.

## Approach

The approach to problem-solving illustrated in this book can be summarized as follows:

- Abstract the problem into a mathematical form with the help of logic and by invoking the appropriate physics principle expressed mathematically.

- Use mathematics to solve for the desired parameter.

- Check the physicalness of the final expression.

- De-abstract the answer into words.

These general steps are applied in the chapter on word problems and each of the main classes of mechanics problems covered in a first-year course on mechanics: Newton's Second Law, Work-Kinetic Energy Theorem, Conservation of Energy, Conservation of Momentum, and Rotational Dynamics and Angular Momentum.

Each chapter has two parts. The first is a discussion of the key principle and how it is applied to solve problems. This section includes a list of the targets and goals along with the skills needed for problem-solving using that principle. Following this is a description of the nature of the types of problems being reviewed along with caveats concerning aspects of using that principle and common pitfalls. The discussion section of the chapter ends with an outline of the basic problem-solving steps and a single-page summary of the steps.

The second part of the chapters are examples of solved problems in a two-column format. The left-hand column has the solution steps printed in a font that resembles a student's writing. The right-hand column has annotations in a more formal font describing what was done in each step with reference to the basic problem-solving steps outlined at the end of the discussion section so that the student knows why a step was done.

Throughout the book, the reader is reminded that although the solutions provided are likely to have more detail than necessary to satisfy a grader, that showing one's work is an essential part of good problem-solving. The reasons are multiple: The more work shown that is unambiguous, the better the grade will be on work that is graded for partial credit. But also, the more work that is shown, the more likely the person creating the solution will be able to catch their own errors.

---

1 For more information on the different types of mathematical understanding students use and how this affects their problem-solving see Skemp, Richard R. "Relational Understanding and Instrumental Understanding." The Arithmetic Teacher 26, no. 3 (1978): 9–15. http://www.jstor.org/stable/41187667.

**EXAMPLE 1.3**

A stone is thrown at a speed of 3.5 m/s from the top of a vertical cliff that is 5.0 m tall at an angle of 35° above the horizontal. How far away in the horizontal direction does the stone land?

| SOLUTION: | ANNOTATION/COMMENT (STEP #): |
|---|---|
| | **Abstracting phrases into mathematical expressions and drawings (1)** |
| 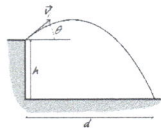 | Figure drawn to identify key parameters and expected behavior. |
| $v_i = 3.5$ m/s, $g = 9.8$ m/s, $\theta = 35°$, $h = 5.0$ m | Writing out the given raw data and any data that can be inferred from the problem in mathematical form. |
| $d = ?$ | Writing out question being asked in mathematical form. |
| $y = \tan\theta \; x - \dfrac{g\,x^2}{2v_i^2 \cos^2\theta}$ | General mathematical relationship that is appropriate for projectile motion. |
| $(x = 0, y = 0)$ at launch point | Writing down reminder of how properly to apply the general principle or relationship being used along with an updated diagram. |
| $\therefore \; y_f = -h, \; x_f = +d$ | "∴" means "therefore" |
| 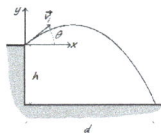 | Note: Writing down reminders for how a relationship must be used is not always necessary, but it is a good idea when starting out to help prevent errors. |
| | Diagram redrawn with coordinate system for 2D trajectory relationship included. |

**TABLE 0.1**  Types of Annotations and their Formatting

**Basic steps as outlined in the Problem-Solving Steps are in bold with the step number in parentheses (#)**

Key results of the application of a basic step are underlined.

Comments on steps specific to the problem being solved have no formatting.

> Note: Notes that identify lessons or points that are applicable generally to all problem-solving are in a green box and identified by the word "Note:". If the corresponding step is in a gray box, it can be considered optional to write down but should at least be done by inspection, (i.e., "in one's head").

The solutions are written using symbolic algebra to demonstrate how this helps check if the solution being created is likely correct. This is shown in three ways:

- First, using symbolic algebra in conjunction with diagrams, logic, and the data given makes it easy to see if there is enough information to solve the problem (e.g., does the number of equations equal the number of unknowns?).

- Second, checking units of symbolic expressions is an easy way to expose any errors that may have occurred, and if an error is found, there is no use going forward with the rest of the solution until the mistake is fixed.

- Third, having a symbolic form for an expression makes it possible to check its "physicalness." This is done by checking its predicted behavior in different limiting cases against the expectations of how the system will respond.

By using this structure for the example solutions, any ambiguity regarding why a certain step was used is hopefully eliminated. And by demonstrating that all problems using the same principle are analyzed and started the same way, students will realize that having learned a general approach to a type of problem, they can now solve any problem of a similar type.

## Topics Covered

This book is focused on the types of problems students encounter in a first-year course in calculus-based mechanics. Oftentimes, such a class is the first in which students are asked to combine subject matter knowledge with mathematical tools. This can be a big conceptual leap and one for which students might not have had much preparation. As such, it is important that they create a strong set of problem-solving skills as these will be the foundation of their problem-solving throughout their STEM courses and later careers.

Accordingly, Chapter 1 is on the topic of word problems, with examples from 1D kinematics. Including this chapter on word problems and putting it first was intentional because every problem in science and engineering is first posed using words. It is the translation of words into a mathematical form that is a challenge for many students to learn.

Throughout the book, this initial translation step is called the "abstraction" step, in which the logical relationships in the sentences are transformed into more mathematical forms. It is the abstraction of ideas into the highly structured, symbolic logic of mathematics that allows us to then bring to bear the powerful tools of algebra and calculus. After the application of our mathematical tools, normally in the form of solving for a variable that represents the value of some parameter, we then "de-abstract" the final mathematical expression back into words. It is the abstraction, use of mathematics, and then de-abstraction that students learn are the main steps of solving any problem they will confront.

Other important steps in the process of solving word problems are also demonstrated, such as checking for logical self-consistency, enough information to solve the problem is available (number of equations equals the number of unknowns), and the logicalness of the final answers.

Chapters 2 through 6 are in the same order as is often found in standard first-year physics textbooks: Newton's Second Law, Work-Kinetic Energy Theorem, Conservation of Energy, Conservation of Momentum, and Rotational Dynamics and Angular Momentum. The chapters were written with the intention of having them mostly independent of the previous ones. Of course, to understand work and then energy, an understanding of forces is required. However, if a student needs help with applying the work-kinetic energy theorem, they will likely only be looking for help after having covered the topic of forces in class so that, in practice, it is likely that these chapters could be used independently of each other.

Lastly, the appendix contains a review of unit analysis and unit conversion, topics often underappreciated by students. The ability to convert units from one type to another is a skill required by any STEM student and is covered in the second half of the appendix. More important is how to use unit analysis as an intermediate check on one's progress while creating a solution; in other words, if a unit analysis check shows that the expression has the wrong units, then a mistake was made that needs to be fixed before moving on.

There are no chapters on the other commonly covered topics of mechanics such as statics, gravitation, oscillations, etc. By the time a student progresses to those topics, they most likely will be able to apply the key principles to those new topics and have developed their problem-solving skills to the point that they can adapt what they have learned to any new topics.

## How to Use This Book

It is not unexpected that students will search through the book looking for problems that are similar to the ones they want to solve instead of reading the introductory discussion on each type of problem. Even if this is done, students will discover in the examples that for each family of physics problem, there is a set of easy-to-follow basic steps that will guide them to solve any problem using the same principle.

A suggested way to use these solutions is for students to cover the solutions and annotations for a problem and try to solve it on their own with only the summary list of the key steps available. Then if they need help, only revealing the next step or annotation to provide a clue as to what to do next. By doing so, they will challenge themselves just enough to reinforce the lesson and avoid the trap of thinking that after only seeing a solution in full, that they could solve a similar problem from scratch.

## Final Thoughts

This book was written to empower students and help them understand that the process of solving physics problems, or any problem for that matter, is understandable and doable, even if they do not yet know what the solution will be. The goal has been to demonstrate how the application of only a few key physics principles combined with logic is necessary to create a mathematical model of a system of interacting objects. After graduation, students will be asked to solve problems that have yet to be solved; The ability to apply a logical problem-solving strategy is what they will need. This book was written in the hope that it will help them do that for any and all problems they may encounter.

# Solving Word and 1D Kinematics Problems

**1**

## Introduction

This chapter is about learning how to solve word problems motivated by the topic of kinematics. For many students, word problems can be quite intimidating. This chapter has been written to help you realize that with a little guidance and structure to your thinking, you can master word problems and not fear them.

*All problems begin as word problems*. No matter which of the science, technology, engineering, or math (STEM) fields you study or enter into for your career, you will be doing word problems. Learning a logical approach to deconstructing the words into mathematical form, applying mathematical tools, and then constructing an answer will serve you well in every STEM field. Hence, the importance of learning this skill.

What does it take to solve word problems? Actually, not that much. It takes a bit of knowledge of how to read them, a bit of structure for the approach, and some practice. Focus on the process and know that it is not necessary that you see the path to the solution immediately; it will become clearer to you as you start the process of solving the problem. You will know you are getting good at it when you find you are more easily correcting your own errors early in the process because you used a logical approach to understanding the problem.

### Skills and Knowledge Needed to Solve Word Problems

◊ *Algebra skills:* The ability to combine and rearrange a system of equations using only symbols to represent values. The ability to solve a system of equations, for example, two equations with two unknowns, will be very useful.

◊ *Basic calculus*: The ability to use derivatives to find the maximum and minimum of an expression. (Not all students will have this knowledge when taking first-year physics.)

◊ *Critical thinking*: The ability to apply simple logic to evaluate constraints and answers.

## TARGETS AND GOALS

In this chapter, you will learn the following:

✔ **How to abstract the words and question to be answered into mathematical forms that can be easily manipulated using mathematical tools, for example, algebra.**

✔ **How to de-abstract the final mathematical relationship(s) back into words and concepts and test final answers against expectations.**

## The Nature of Word Problems

Know that whatever problem you may be attempting to solve, it is a process you can master. Even if it feels that the way to answer a problem just popped into your head, know that your mind was processing the information while you were reading it. **This chapter (and book) is about elevating your awareness of the process of problem-solving so that all problems will become easier to solve.** This is done by providing a discussion of examples of combining and leveraging your intuitive problem-solving processing with a bit more structure and some mathematical tools to increase the range of problems that you will be able to solve. By becoming even a little more aware of the process by which you solve problems, your ability to solve any problem, whether it be in your physics courses or another STEM-related field, will be greatly enhanced.

It is worth repeating that when starting a problem, the path to answer or even the question being asked will not always be obvious. The act of de-abstracting the problem, drawing pictures to describe what is expected to happen, and writing down any other mathematical expressions that seem appropriate will result in new insights into how to proceed. In short, it is the process of problem-solving that you will be learning. It is in the application of the process that the path to the answer will reveal itself.

One of my favorite expressions that I will repeat several times in this book is, "*A well-understood problem is a problem half-solved.*" This is the first and most critical step in solving any problem, consisting of revealing the nature of the problem itself.

**This initial analyzing and reframing of the problem into new words and mathematical expressions is what working engineers and scientists do to figure out how to solve problems no one has solved before. Once this has been done, what to do next will be much more obvious.**

**The Basic Steps to Solving Word Problems:**
1. *Abstract* the words into mathematical and/or graphical restatements of the words and identify any additional needed mathematical relationships.
2. *Apply mathematical tools* such as algebra to manipulate the mathematical expressions as necessary to create an expression for the desired quantity.
3. *De-abstract* the final mathematical expression(s) into words that describe the relationship found and check if it makes sense.

Even more succinctly: *Solving word problems is an act of translating logical relationships expressed in words into logical mathematical statements that then are rearranged as necessary so that a logical conclusion can be drawn from a final mathematical expression.*

Along the way, we will identify specific ways to turn what is in words into equations, even ones that are not explicitly shown, and how to apply logical constraints to what the answers or parameters could be for the answer to be reasonable.

## The Problem-Solving Steps for Word Problems

Below is a description of the basic steps for solving a word problem, written out so that you can understand the approach. After this overview, we will apply these ideas in a more structured manner with annotated examples.

### Approach

For word problems, the adage, "A well understood problem is a problem half-solved" is put into practice by taking the whole (sometimes apparently complicated) problem and breaking it down into smaller parts or sub-problems. This would contrast with blindly applying an algorithm and not knowing why it was the appropriate one to use.

These are key questions to ask yourself when starting your solution:

- What are the key parameters, both explicit and implicit, in the problem statement?

- What relationship(s) will be needed, when combined with the data provided, to answer the question being asked?

- Is there more than one part to the problem that must be solved before another?

### Identify the Type of Problem

Generally, every problem is a word problem, but not all are necessarily also a physics problem. Non-physics word problems ask for the consequences of a logical set of constraints that do not require the additional use of a physics principle to find the answer, for example, "What temperature has the same numerical value in Fahrenheit and Celsius?" This definition is not intended to compartmentalize the different types of problems as being totally different from each other but rather as a way to identify those that will and will not also need some physics principle to complete.

For the purposes of this book, simple kinematics problems will be used to demonstrate how word problems are solved. Being kinematics problems, these will require the additional step of identifying the kinematics relationship(s) necessary to solve the problem. The skills developed on these problems in the abstraction of the question will be needed and used for all other physics, math, and engineering problems you will encounter.

## Hallmarks of a Word Problem

Words are used to describe a situation from which predictions are to be made or conclusions drawn based on the parameters and/or logical relationships provided or that can be inferred that describe the elements of the system.

## Abstract the Problem

This is the key step in the solving of word problems. The goal is to glean as much information from the problem statement and translate it into a mathematical form. Once it is in a mathematical form, then the power of mathematics is used to rearrange the ideas in symbolic form, which is much easier than if we had been limited to using only full written sentences.

A few examples of how to do this translation would be:

- Stating definitions of constants:
    - "*The initial height is 10.0 meters.*" ↔ "$h_{\text{init}} = 10.0$ m"
    - "*The car came to rest.*" ↔ "$v_{\text{car, final}} = 0$"
- Restating a provided or inferred relationship:
    - "*The height above the ground can only be a positive value.*" ↔ "$h \overset{!}{\geq} 0$".[1]
    - "*When the police car catches up to the speeder*" ↔ "$x_p(t^*) = x_s(t^*)$"
- Expressing the main question as an equation:
    - "*What was the car's final velocity?*" ↔ "$v_f = ?$"

Sometimes it is not possible to anticipate all of the equations that will be needed to solve the question at hand. That is OK. You will fill in more as needed and as you work the problem.

Once the data and information about the problem has been re-expressed in mathematical form, the next step is to identify any additional mathematical relationships that would be required to answer the question. For a physics problem, and specifically one on kinematics, this would mean identifying the relationship that would relate the given information to that needing to be found, for example, $x_f = x_i + v_i t + \frac{1}{2}at^2$.

## Applying Mathematics

After the problem has been translated into an initial mathematical form, algebra and/or calculus is used to solve for the unknown variable(s). Although, in principle, it may be possible to do simple word problems using logical tools that are written out in words or even "in one's head," in practice, this non-mathematical approach is clumsy and limited in its capabilities. To overcome these drawbacks, the highly structured, symbolic logic of algebra and mathematics is used. For the small cost of abstracting the problem into symbolic form, many powerful mathematical tools for rearranging the mathematical relationships to solve for the desired answer become available.

An important idea to keep in one's mind when doing mathematical manipulations is that if your starting mathematical relationships are correct and your algebra steps are correct, whatever your final result is, it too will be correct. The beauty of algebra is that you can do the mathematical steps in a different order, and the answer will always be the same. The other important idea to keep in mind is that mathematical manipulations (algebra) are not calculations. Calculations will occur in the final step.

As you do your mathematical steps, your goal should be to do all of them in a way that is as unambiguous as possible to anyone who might look at your work (including yourself). This is best accomplished by doing all of your algebra symbolically and writing out all your steps. Doing so ensures that you minimize any algebra errors and making it easier to check your work. Also, it has the benefit that anyone grading your work will have an easier time finding evidence that you know what you are doing and give you an accurate grade. Thus, resist the temptation to minimize the amount of paper you use for writing out your solutions. Trying to do too many steps in your head will only result in more mistakes, greater difficulty in solving the problem, and in the end, more time to solve the problem. If necessary, once you are done with a solution, you can copy it so that the workflow is more obvious to those who will be evaluating it.

---

1   The exclamation mark above the inequality means "must." In other words, the distance must be greater than or equal to zero.

Part of doing your algebra will be checking if the number of equations you have equals the number of unknowns (variables). If it does, you likely have a solvable system of equations.[2] When solving a system of equations, it may seem intimidating to face lots of algebra without knowing if the work ahead will result in the desired answer. However, having the same number of equations as unknowns should give you confidence that you are on track to finding a solution.

Another check to make that will boost your confidence in your equations is to examine the units in the relationships. An example of an equation in which the units do not agree would be this (incorrect) one for distance: $D = v^2 t$. Using square brackets $[x]$ to represent "*units of x*", $[v^2 t] = (m/s)^2 \cdot s = m^2/s$, which does not agree with the expected units for distance of meters. If you find such a discrepancy in your units, go back, and check the starting relationships and algebra before continuing. This will save you much time later.

Regarding the use of calculus, consider its use as just another set of tools in your mathematical toolbox. The simplest examples will be the use of derivatives to find the maximum or minimum of functions.

Finally, do not hesitate to be explicit with notation. For instance, when solving for a special value of a variable, say the time at which an object reaches a certain distance, we will represent the variable for time with a prime or an asterisk as a superscript or something else as a subscript: $x(t') = v_0 t' = x_{stop}$. The translation of this equation would be "*At the particular time $t'$ the position of the object given by the general expression $x(t)$ evaluated at $t'$ has the same value as $x_{stop}$.*" The use of the prime designates this as a specific value of time instead of the independent variable that is used in the general expression for the position as a function of time.

A calculus example would be the following: "*The specific value(s) of x for which the function y of x is at a maximum or minimum is defined as $x^*$*" would be written as "$\left. \dfrac{dy(x)}{dx} \right|_{x^*} = 0$." Here, the vertical bar with $x^*$ at its base indicates that $x^*$ is the value of $x$ that solves this the equation.

Finally, a word of caution here about when to insert numerical values into problems:

Resist the urge to insert numerical values into the mathematical expression for the problem before the final steps and especially before doing any algebra. The reason is that it will be easier to check your own work, and in some cases, there will be an unknown value that can be eliminated or canceled that is not provided. Furthermore, in more advanced physics and engineering courses, these expressions will be used to calculate a range of values for several different sets of initial parameters. If the problem is not solved symbolically first, then for each new set of parameters, the algebra will have to be repeated for each new calculation.

## De-Abstracting the Answer

"De-abstracting" is done after the mathematical steps. This is the final step of translating the final result back into words to answer the original question. Part of the process of this de-abstraction is to check that the answer makes logical and/or physical sense. For instance, a final speed greater than the speed of light is not physical (unless you are considering science-fiction stories). For that case, this would be a clue that somewhere a mistake was made or that what was desired is impossible to accomplish. This is where the logic of the de-abstraction step is applied to forming your final answer.

## Tools for Solving Word-Problems

On the following pages are two tables that show examples of words translated into mathematical expressions and drawings, and mathematical expressions and drawings translated into words. These lists are not exhaustive but should give you a good idea on how to do the translations.

Following the tables is a list of the basic steps in solving a generic word problem. Considering that there are many possibilities for solving such problems, this list should be considered a guide rather than a rigid algorithm to follow. The goal is to be able to answer the question, "*How did I know to do that step?*" In the chapters on solving force, energy, momentum, and other problems, the steps can be followed more closely.

---

2   There are cases when one of your equations is a rearrangement of one of the others. In such cases, these equations are only alternate versions of the same relationship (i.e., they are not linearly independent). If this is the case, then one more equation or more data will be needed to solve the problem.

TABLE 1.1 Translating Words to Mathematical Expressions

| Written Expression | Mathematical and Graphical Translation |
|---|---|
| "The car started from rest." | $v_{car,i} = 0$ |
| "The block came to rest." | $v_{block,f} = 0$ |
| "The car is accelerating backwards at 3.2 m/s²." | $a_{car} = -3.2 \ \text{m/s}^2$ |
| "An object is released from rest and falls." | $v_i = 0, a = -g$ (for y-axis pointing up) |
| "When does car 2 overtake car 1?" "At what distance down the road does car 2 overtake car 1?" | $t^* = ?$ for $x_1(t^*) = x_2(t^*)$ <br> $x_1(t^*) = x_2(t^*) = x^* = ?$ <br> |
| "What is the final speed of the object?" | $v_f = ?$ |
| "How fast is plane 1 traveling when it is overtaken by plane 2?" | $v_1 = ?$ when $x_1(t^*) = x_2(t^*) = x^*$ |
| "The time it takes for ball 1 to fall is three times the time it takes ball 2 to fall." | $t_1 = 3t_2, t_1/t_2 = 3$ |
| "The time for rock 2 to fall is 2 seconds less than that for rock 1." "2.0 seconds after the first rock is dropped, the second one is dropped." | $t_2 = t_1 - t_0, t_0 = 2.0 \ \text{s}$ |
| "The speed of the car moving backwards is 25 m/s." | $\|v_{car}\| = \|-25 \ \text{m/s}\| = 25 \ \text{m/s}$ <br> $\|\vec{v}_{car}\| = 25 \ \text{m/s}$ <br><br> Note: The absolute value signs can be used on the variable for the velocity if it is a vector or a scalar. |
| "The initial velocity is 200 m/s in the direction 30 degrees north of east." | $v_i = 200 \, \text{m/s}, \theta = 30°$ <br> $\vec{v}_i = v_i \cos\theta \, \hat{i} + v_i \sin\theta \, \hat{j}, \theta = 30°$ <br> $= 173 \ \text{m/s} \, \hat{i} + 100 \ \text{m/s} \, \hat{j}$ <br> North $= \hat{j}$, East $= \hat{i}$ <br> |

TABLE 1.2  Translating Mathematical Expressions into Words

| Mathematical and Graphical Expression | Written Translation |
|---|---|
| $$y_f \overset{!}{\geq} 0$$ | "The final height must be greater than zero." |
| | "The final height above the ground cannot be a negative value (for $y = 0$ being ground level)." |
| $$v_f^2 \overset{!}{\geq} 0$$ | "The square of the final speed must be positive to be physical." |
| $(b^2 - 4ac) \overset{!}{>} 0$ or $\sqrt{b^2 - 4ac} \overset{!}{\in} \mathbb{R}$ <br> Note: These situations can arise when there are square root signs in an expression or when a square root will be taken. | "The argument of the square root function must be positive." or "The value of the square-root must be real.' |
| $$v_f = \pm 2.0 \text{ m/s}$$ <br> Note: This can arise when a square root of a squared value is encountered. E.g., $v_f^2 = 4.0 \text{ m}^2/\text{s}^2 \Leftrightarrow v_f = \pm 2.0 \text{ m/s}$ | "The final velocity is plus or minus 2.0 m/s." |

# Word Problems Problem-Solving Steps

<u>Hallmarks:</u> Words are used to describe a situation from which predictions are to be made or conclusions drawn based on the parameters and/or logical relationships provided or that can be inferred that describe the elements of the system.

## Steps

1. **Abstract the word problem into mathematical form\***
   - *Draw a figure or diagram*
     - to describe expected behavior
     - to identify coordinate systems
   - *Translate the words into mathematical relationships*
     - Create mathematical expressions for
       - the given values
       - the values desired/the question to be answered
   - *Identify additional relationship(s) that are needed*[3]
     - logical statements that can be inferred from the problem statement (e.g., constraints on the values for any variables or answers)
     - kinematics equations that would be helpful to answer the question
   - *Check if there is enough information*
     - Check if the number of equations equals the number of unknowns
       - If Yes, proceed
       - If No, check for other data or logical constraints to include
2. **Apply the mathematics tools to the relationships created**
   - *Use symbolic algebra to solve the system of equations* as necessary to find an expression or relationship that provides the information asked in the problem statement
   - *Check units* of the final expression if necessary
   - *Insert the numerical values* into the final expression[4] and evaluate
3. **De-abstract the answers**
   - *Check physicalness* of predictions of final relationship(s)
   - *De-abstract the final mathematical expression(s)* using brief words or short sentences

(*\*This is the most important step for word-problems.*)

---

3   If not a kinematics problem, this step would include identifying any mathematical relationships or definitions that would be helpful in answering the question. For example, for the question, "*What temperature has the same numerical value in Fahrenheit and Celsius?*" the relationship to use would be $T_C = (5/9)(T_F - 32)$.

4   There can be instances when it is practical to compute an intermediate value for an expression rather than wait until the final steps of a problem. This will occur when there is part of an expression that shows up in many places. In general, it is advisable to limit doing this so that any final expressions will be in the easier-to-check symbolic form.

# EXAMPLE 1.1

A cart is set rolling up from the bottom of a ramp at a speed of 4.0 m/s. The acceleration it experiences is 3.2 m/s² directed down the ramp.

a. What is its position along the ramp relative to its initial position when its velocity is 1.0 m/s going forward?

b. What is its position along the ramp relative to its initial position when its velocity is 1.0 m/s going back down?

| SOLUTION: | ANNOTATION/COMMENT (STEP #): |
|---|---|
| | **Abstracting phrases into mathematical expressions and drawings (1)** |
| 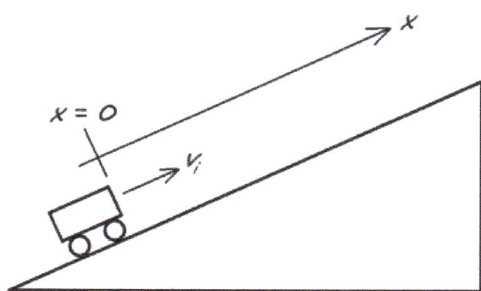 | The picture helps to identify the coordinate system to use and its origin. |
| $v_i = +4.0 \text{ m/s}, \ a = -3.2 \text{ m/s}^2$ <br> $x_i = 0$ | Writing out the data given and that can be inferred in mathematical form. |
| | Here, the choice of $x_i$ was free, so for convenience, it was chosen to be zero. |
| $x = ?$ <br> when $v_f = +1.0 \text{ m/s}$ <br> when $v_f = -1.0 \text{ m/s}$ | Writing out question being asked in mathematical form. <br> For this coordinate system, "*going forward*" is the plus-direction and "*going back down*" is the minus-direction. |
| | Note: Good practice is to explicitly include plus (+) and negative (−) signs in all expressions in the data not only the magnitudes of the values. This prevents calculation and algebra errors later. |
| $2a(x_f - x_i) = v_f^2 - v_i^2$ | Mathematical expression of the logical or physical situation being examined. |
| | Note: Whenever possible, try to identify the equation to use as the one for which all the variables were given except for the one of interest. |

| SOLUTION: | ANNOTATION/COMMENT (STEP #): |
|---|---|
| $$2a(x_f - 0) = v_f^2 - v_i^2$$ $$2ax_f = v_f^2 - v_i^2$$ $$x_f = \frac{v_f^2 - v_i^2}{2a}$$ | **Using mathematics to solve for the value of interest (2)** |
| | Note: Solving expressions symbolically reduces the number of algebra errors, makes checking the steps easier, and gives a final expression that can be used for different input values. |
| a) $x_f = \dfrac{v_f^2 - v_i^2}{2a}$ $$= \frac{(+1.0 \text{ m/s})^2 - (+4.0 \text{ m/s})^2}{2(-3.2 \text{ m/s}^2)}$$ $$= \frac{2.3 \text{ m}^2/s^2}{\text{m}/s^2}$$ | Note: Units are included in the evaluation so that the final answer can be checked. If there is an error with units, then there was a mistake to be fixed. |
| $\boxed{x_f = 2.3 \text{ m}}$ | Value for final position for $v_f$ = +1.0 m/s. |
| b) $x_f = \dfrac{v_f^2 - v_i^2}{2a}$ $$= \frac{(-1.0 \text{ m/s})^2 - (+4.0 \text{ m/s})^2}{2(-3.2 \text{ m/s}^2)}$$ | Repeating calculation but for $v_f$ = −1.0 m/s. |
| $\boxed{x_f = 2.3 \text{ m}}$ | Value for final position for $v_f$ = −1.0 m/s. |
| "The final positions for $v_f = \pm 1.0$ m/s are $x_f = 2.3$ m." | **De-abstractions of the final mathematical/logical expression (3)** |
| | Final de-abstraction of numerical answers into words. |
| | Here the same speed (not velocity) would be expected on both the way up and down the ramp for the same distance. |

# EXAMPLE 1.2

A stone dropped from rest from the top of a cliff takes 4.5 times longer to hit the ground than when a second stone is dropped off another cliff. Ignoring the effects of air drag, find the height of the second cliff in terms of the height of the first cliff.

**SOLUTION:**

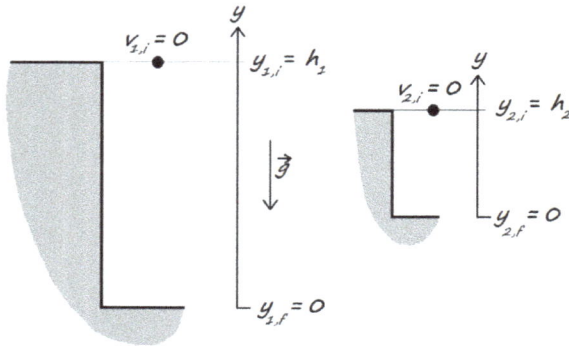

**ANNOTATION/COMMENT (STEP #):**

**Abstracting phrases into mathematical expressions and drawings (1)**

The key parameters are inserted into the picture to help identify all given values and those that are only inferable from the context.

Here the bottom of the cliff is set to be $y = 0$ with the $y$-axis pointing up.

$v_{1,i} = 0, v_{2,i} = 0$

$a_1 = -g, a_2 = -g$

$y_{1,j} = h_1, y_{2,j} = h_2$

Writing out the given raw data and any data that can be inferred from the problem in mathematical form.

Inferred $a = -g$ and $v_i = 0$ from "dropped" and "dropped from rest" along with the direction of the chosen coordinate system.

$t_1 = 4.5t_2$

The time it takes for the first rock to fall is 4.5 times longer than the for the second.

$\dfrac{h_2}{h_1} = ?$

Writing out question being asked in mathematical form.

$y_f = y_i + v_i t + \frac{1}{2}at^2$

General mathematical expression that is appropriate for free-fall as a function of time.

**Using mathematics to solve for the value of interest (2)**

Adapting general key mathematical relationship to the first and second instance and simplifying.

$y_{1,f} = y_{1,i} + v_{1,i}t + \frac{1}{2}a_1 t_1^2$

$0 = h_1 + 0 + \frac{1}{2}(-g)t_1^2$

$\boxed{h_1 = \frac{1}{2}gt_1^2}$

## SOLUTION:

$$y_{2,f} = y_{2,i} + v_{2,i}t + \frac{1}{2}a_2 t_2^2$$

$$0 = h_2 + 0 + \frac{1}{2}(-g)t_2^2$$

$$\boxed{h_2 = \frac{1}{2}gt_2^2}$$

$$\frac{h_1}{h_2} = \frac{\left(\frac{1}{2}gt_1^2\right)}{\left(\frac{1}{2}gt_2^2\right)} = \frac{\cancel{\frac{1}{2}g}\; t_1^2}{\cancel{\frac{1}{2}g}\; t_2^2}$$

$$\frac{h_1}{h_2} = \left(\frac{t_1}{t_2}\right)^2$$

$$\boxed{h_2 = \left(\frac{t_2}{t_1}\right)^2 h_1}$$

$$h_2 = \left(\frac{\cancel{t_2}}{4.5\,\cancel{t_2}}\right)^2 h_1 \quad h_2 = \left(0.222\right)^2 h_1$$

$$\boxed{h_2 = 4.9\% \; h_1}$$

"The height of the second cliff is 4.9% of the first."

## ANNOTATION/COMMENT (STEP #):

<u>Note:</u> All the substitution steps are being shown, including the zeros, to eliminate any ambiguity. It is generally better to include more steps than fewer. Although more is written, it is generally faster. Also, doing so results in fewer mistakes, and if mistakes are made, they are easier to find.

Creating the ratio of the heights to evaluate and then simplifying. In the numerator and denominator, the expressions for $h_1$ and $h_2$ are substituted.

<u>Note:</u> To keep track of where a substitution was done, parentheses are used. Doing this can make checking the algebra easier.

<u>Final algebraic expression for the height of the second cliff.</u>

<u>Numerically evaluating final expression.</u>

**De-abstraction of the final mathematical/logical expression (3)**

This is not unreasonable because it would be expected that less time would be needed to fall to the bottom of a shorter cliff.

# EXAMPLE 1.3

A stone is thrown at a speed of 3.5 m/s from the top of a vertical cliff that is 5.0 m tall at an angle of 35° above the horizontal. How far away in the horizontal direction does the stone land?

**SOLUTION:**

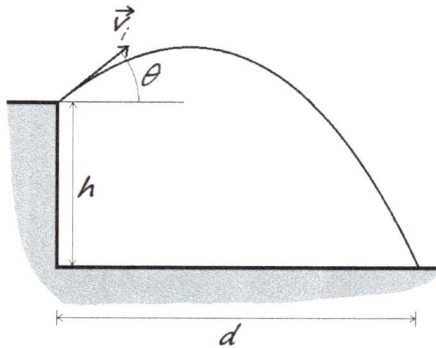

**ANNOTATION/COMMENT (STEP #):**

**Abstracting phrases into mathematical expressions and drawings (1)**

Figure drawn to identify key parameters and expected behavior.

$v_i = 3.5$ m/s, $g = 9.8$ m/s, $\theta = 35°$,
$h = 5.0$ m

Writing out the given raw data and any data that can be inferred from the problem in mathematical form.

$d = ?$

Writing out question being asked in mathematical form.

$$y = \tan\theta\, x - \frac{g\,x^2}{2v_i^2 \cos^2\theta}$$

$(x = 0, y = 0)$ at launch point

General mathematical relationship that is appropriate for projectile motion.

Writing down reminder of how properly to apply the general principle or relationship being used along with an updated diagram.

$\therefore\ y_f = -h,\ x_f = +d$

"$\therefore$" means "therefore"

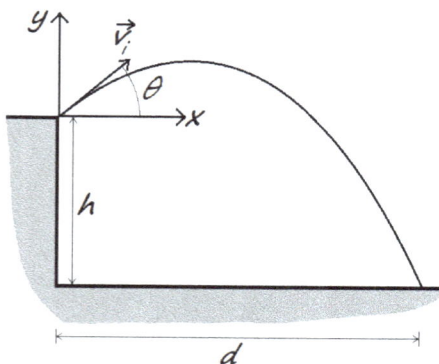

Note: Writing down reminders for how a relationship must be used is not always necessary, but it is a good idea when starting out to help prevent errors.

Diagram redrawn with coordinate system for 2D trajectory relationship included.

| SOLUTION: | ANNOTATION/COMMENT (STEP #): |
|---|---|

### SOLUTION:

$$(-h) = \tan\theta\,(d) - \frac{g(d)^2}{2v_i^2\cos^2\theta}$$

$$\left(\frac{g}{2v_i^2\cos^2\theta}\right)d^2 + (-\tan\theta)d + (-h) = 0$$

$$\left(\frac{g}{2v_i^2\cos^2\theta}\right) = \frac{9.8\ m/s^2}{2(3.5\ m/s)^2\cos^2 35^0}$$

$$= 0.596\left(\frac{\cancel{m}/\cancel{s^2}}{m^{\cancel{2}}/\cancel{s^2}}\right)$$

$$\left(\frac{g}{2v_i^2\cos^2\theta}\right) = \frac{0.596}{m}$$

$$\tan 35^0 = 0.700$$

$$\left(\frac{0.596}{m}\right)d^2 + (-0.700)d + (-5.0m) = 0$$

$$d = \frac{1}{2(0.596/m)}\left[0.700 \pm \sqrt{(0.700)^2 - 4\left(\frac{0.596}{\cancel{m}}\right)(-5.0\ \cancel{m})}\right]$$

$$= (0.839\ m)(0.700 \pm 3.52)$$

$$= (-2.37\ m, +3.54\ m)$$

$$d \overset{!}{\geq} 0 \quad \therefore\quad d = 3.54\ m$$

"The stone lands at x = 3.54 m."

### ANNOTATION/COMMENT (STEP #):

**Using mathematics to solve for the value of interest (2)**

The appropriate symbolic values are substituted, then rearranged into the form to use the quadratic formula.

Note: Parentheses around substitutions can help keep track of where they are inserted.

Calculating the coefficients of the polynomial.

Note: Even when using the quadratic formula, units matter. Notice how they cancel and propagate to give the expected units. Checking units helps to check if the calculations are being done properly.

Calculating numerical coefficients for use in the quadratic formula:

$$ax^2 + bx + c = 0,\quad x = \frac{-b \pm \sqrt{b^2 - 4ac}}{2a}$$

**De-abstractions of the final mathematical/logical expression (3)**

Numerical calculations using quadratic formula.

The units and minus signs cancel so that the value of the square-root will be real and that the answer has the correct units.

The quadratic formula always provides two answers. Here only $d > 0$ makes sense. The "!" above the $\geq$ sign is read as "must be".

Applying logic to final answer. Here the stone can only land out front of the cliff.

# EXAMPLE 1.4

Jill leaves a rest-stop 5 minutes after her friend Fred hoping to catch up to him before his exit 40 miles ahead. Fred drives at a constant speed of 45 miles per hour. At what average speed must Jill drive to catch Fred before his exit? What would her speed need to be if she left 10 minutes after Fred?

**SOLUTION:**

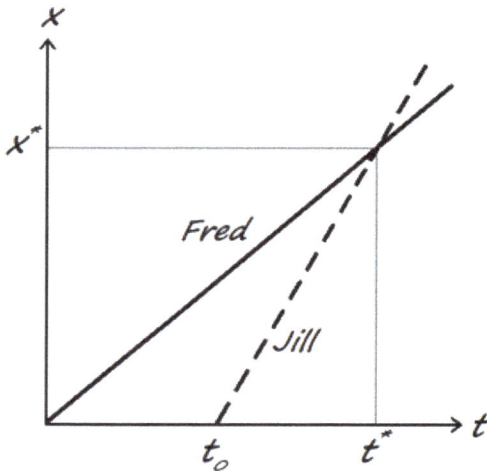

**ANNOTATION/COMMENT (STEP #):**

**Abstracting phrases into mathematical expressions and drawings (1)**

Figure drawn to identify all key parameters and show expected behavior.

Because speed is constant, the lines of $x$ versus $t$ are straight. Jill starts at a later time that is labeled $t_0$.

$$v_{F,i} = 45 \text{ m/s}, \; a_F = 0, \; a_J = 0,$$

$$t_o = 5 \text{ min.}, 10 \text{ min.}$$

$$x^* = 40 \text{ miles}$$

$$x_{F,i} = x_{J,i} = 0$$

Writing out the given raw data and any data that can be inferred from the problem in mathematical form.

The choice of the value at the starting point was free to be made. To simplify later calculations, it was chosen to be zero.

$$v_J = ? \text{ when } x_F(t^*) = x_J(t^*) = x^*$$

Writing out question being asked in mathematical form.

The position where and the time when Jill catches Fred are labeled $x^*$ and $t^*$ because they are specific values.

$$x_f = x_i + v_i t + \tfrac{1}{2} a t^2$$

General mathematical relationship that is appropriate for this motion.

This expression will be adapted to both Jill's and Fred's situation.

| SOLUTION: | ANNOTATION/COMMENT (STEP #): |
|---|---|
| $$x_F = x_{F,i} + v_{F,i}t + \tfrac{1}{2}a_F t^2$$ $$x_F = 0 + v_{F,i}t + 0$$ $$x_F = v_{F,i}t$$ | **Using mathematics to solve for the value of interest (2)** Inserting zero values in this step to simplify things for later. |
| $$x_J = x_{J,i} + v_{J,i}(t - t_o) + \tfrac{1}{2}a_F(t - t_o)^2$$ $$x_J = 0 + v_{J,i}(t - t_o) + 0$$ $$x_J = v_{J,i}(t - t_o)$$ | The time that Jill travels is shorter than Fred's by the amount $t_0$. This is why the correct expression to use for the time in the equation of motion for Jill is $(t - t_0)$ and not $t$. |
| <u>Final equations:</u> $$x_F(t^*) = x_J(t^*), \quad x_F(t^*) = x^*$$ $$x_F = v_{F,i}t^*, \quad x_J = v_{J,i}(t^* - t_o)$$ | Consolidating the final equations The first equation is the "equation of constraint," or the "coupling equation," which relates Jill's motion to Fred's. The last two are the expressions for where Fred and Jill will be when Jill passes Fred. |
| equations: 4 unknowns: 4; $t^*, v_{J,i}, x_F, x_J$ knowns: $v_{F,i}, t_o, x^*$ | Checking that the number of equations equals the number of unknown values. Note: Listing the number of equations and number of unknowns can often be done in one's head. However, for complicated problems, it can be extremely useful to list them. |
| $$x_F(t^*) = x^*$$ $$v_F t^* = x^*$$ $$t^* = \frac{x^*}{v_F}$$ | Evaluating the relationship for Fred and rearranging to solve for $t^*$, the time when Jill passes Fred, so that it can be substituted into the next expression. This step was chosen to do first because it is such a simple expression to rearrange. The use of $x_F(t^*)$ is used to specify that this is being evaluated at a specific time and is not to be treated as a general relationship. |

**SOLUTION:**

**ANNOTATION/COMMENT (STEP #):**

$$x_F(t^*) = x_J(t^*)$$
$$\Downarrow \qquad\qquad \Downarrow$$
$$(x^*) = (v_{J,i}(t^* - t_o))$$
$$v_{J,i} = \frac{x^*}{(t^* - t_o)}$$

Solving the equation that represents Jill and Fred being at the same place at the same time.

On the left $x_F(t^*)$ was substituted with $x^*$, and on the right side, the expression for $x_J(t)$ evaluated at $t^*$ was substituted. Parentheses are used to show where substitutions were done.

$$v_{J,i} = \frac{x^*}{((x^*/v_F) - t_o)}$$

Final expression for the speed Jill must travel.

$$v_{J,i} = \frac{40 \text{ mi}}{\left( \dfrac{40 \text{ mi}}{45 \text{ mi/hr}} - t_o \right)}$$
$$= \frac{40 \text{ mi}}{\left( \frac{8}{9} \text{ hr} - t_o \right)}$$

Performing preliminary numerical evaluation of the expression for Jill's speed.

Note: Doing a preliminary numerical calculation can make doing the calculations for several values easier.

Note: Including units and then canceling them is a good way to double-check that the algebra was done correctly.

$$v_{J,i} = \frac{40 \text{ mi}}{\left( \frac{8}{9} \text{ hr} - 5 \text{ min} \cdot \left( \dfrac{1 \text{ hr}}{60 \text{ min}} \right) \right)}$$
$$= \frac{40 \text{ mi}}{\frac{8}{9} \text{ hr} - \frac{1}{12} \text{ hr}} = \boxed{50 \text{ mi/hr}}$$

The speed Jill would have to drive if she left 5 minutes late.

$$v_{J,i} = \frac{40 \text{ mi}}{\left( \frac{8}{9} \text{ hr} - 10 \text{ min} \cdot \left( \dfrac{1 \text{ hr}}{60 \text{ min}} \right) \right)}$$
$$= \frac{40 \text{ mi}}{\frac{8}{9} \text{ hr} - \frac{1}{6} \text{ hr}} = \boxed{55 \text{ mi/hr}}$$

The speed Jill would have to drive if she left 10 minutes late.

**De-abstraction of the final mathematical/logical expressions (3)**

"If Jill leaves 5 minutes or 10 minutes late, she would have to travel at 50 or 55 mph, respectively."

Final de-abstraction of the answer

# EXAMPLE 1.5

A car is accelerating at a rate of 1.5 m/s along a curved road with a radius of 400 m. What is the magnitude and direction of its total acceleration when it is traveling at a speed of 15 m/s? What are the magnitude and direction when it is traveling at 20 m/s?

**SOLUTION:**

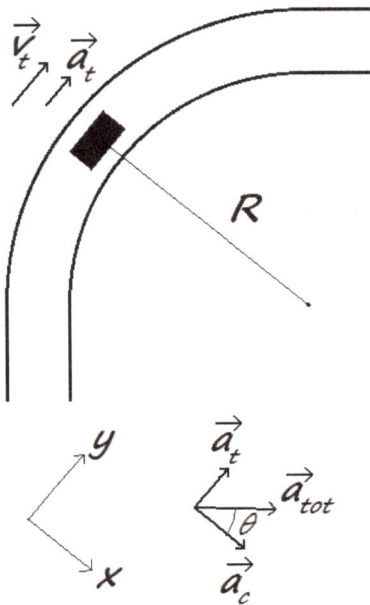

$R = 400 \text{ m}, \ a_t = 0.50 \text{ m/s}^2$

$v_t = 15 \text{ m/s}, \ v_t = 20 \text{ m/s}$

$a_{tot} = ?, \ \theta = ?$

$\vec{a}_{tot} = \vec{a}_c + \vec{a}_t$

$a_{tot}^2 = a_c^2 + a_t^2$

$a_c = v_t^2/r$

$\tan \theta = a_t/a_c$

**ANNOTATION/COMMENT (STEP #):**

**Abstracting phrases into mathematical expressions and drawings (1)**

Draw a picture of what is happening.

Added vectors to picture that will then also be identified below.

Because the car is accelerating in the direction of motion (along its path), $a_{tot}$ will point inward and forward.

Create coordinate system appropriate for the problem to be solved.

Note: For problems with centripetal acceleration, because $\vec{a}_c$ always points inward, a good choice for the x-axis is to make it point inward too.

Writing out the given raw data and any data that can be inferred from the problem in mathematical form.

Writing out question being asked in mathematical form.

Supplementary relationships identified.

These are the relationships that are anticipated that could be useful for solving the problem later.

| SOLUTION: | ANNOTATION/COMMENT (STEP #): |
|---|---|

**Using mathematics to solve for the value of interest (2)**

$$a_{tot}^2 = \left(v_t^2/R\right)^2 + a_t^2$$

$$a_{tot} = \sqrt{v_t^4/R^2 + a_t^2}$$

$$= \left(\frac{v_t^4}{(400\ m)^2} + \left(1.5\ \frac{m}{s^2}\right)\right)^{1/2}$$

$$a_{tot} = \left(\frac{v_t^4}{1600\ m^2} + 2.25\ \frac{m^2}{s^4}\right)^{1/2}$$

Performing substitution for $a_c$ and rearranging to solve for $a_{tot}$.

Note: Creating an expression with one of the data values as a variable allows its use for repeated calculations at the end.

General final expression for $a_{tot}$.

$$\tan\theta = \frac{a_t}{a_c} = \frac{a_t}{(v_t^2/R)} = \frac{a_t\ R}{v_t^2}$$

$$= \frac{(1.5\ m/s^2)(400\ m)}{v_t^2}$$

$$\tan\theta = \frac{600\ m^2/s^2}{v_t^2}$$

Using algebra to construct a general final expression for $\tan\theta$.

General expression for $\tan\theta$.

$$\boxed{a_{tot} = \left(\frac{v_t^4}{1600\ m^2} + 2.25\ \frac{m^2}{s^4}\right)^{1/2}}$$

Partially evaluated final expressions prepared for solving for the two cases requested.

$$\boxed{\tan\theta = \frac{600\ m^2/s^2}{v_t^2}}$$

$$a_{tot} = \left(\frac{\left(15\ m/s\right)^4}{1600\ m^2} + 2.25\ \frac{m^2}{s^4}\right)^{1/2}$$

$$= \left(32.6\ \frac{m^4{\scriptstyle 2}}{m^2\cdot s^4} + 2.25\ \frac{m^2}{s^4}\right)^{1/2}$$

$$= 5.8\ m/s^2$$

Performing numerical evaluation of the expression starting with $v_t = 15$ m/s.

Value of $a_{tot}$ for $v_t = 15$ m/s.

| SOLUTION: | ANNOTATION/COMMENT (STEP #): |
|---|---|

$$a_{tot} = \left[ \frac{(20 \text{ m/s})^4}{1600 \text{ m}^2} + 2.25 \frac{\text{m}^2}{\text{s}^4} \right]^{1/2}$$

Performing numerical evaluation of the expression using $v_t = 20$ m/s.

$$= \left[ 100 \frac{\text{m}^2}{\text{s}^4} + 2.25 \frac{\text{m}^2}{\text{s}^4} \right]^{1/2}$$

$$= 10 \text{ m/s}^2$$

Value of $a_{tot}$ for $v_t = 20$ m/s.

$$\tan\theta = \frac{600 \text{ m}^2/\text{s}^2}{(15 \text{ m/s})^2} = 2.67$$

$$\theta = \tan^{-1}(2.67) = 69°$$

Value of $\theta$ for $v_t = 15$ m/s.

$$\tan\theta = \frac{600 \text{ m}^2/\text{s}^2}{(20 \text{ m/s})^2} = 1.50$$

$$\theta = \tan^{-1}(1.50) = 56°$$

Value of $\theta$ for $v_t = 20$ m/s.

**De-abstraction of the final mathematical/logical expression (3)**

"For $v_t = 15$ m/s, $a_{tot} = 5.8$ m/s$^2$ and $\vec{a}_{tot}$ points forward of directly inward at an angle of 69 degrees.

Final de-abstraction of the mathematical answers into words for each case presented in the problem.

For $v_t = 20$ m/s, $a_{tot} = 10$ m/s$^2$ and $\vec{a}_{tot}$ points forward of directly inward at an angle of 56 degrees."

The total acceleration is expected to be larger when the car is going faster because the centripetal acceleration would be larger. Therefore, the answers make sense.

# 2 Solving Newton's Second Law Problems

## TARGETS AND GOALS

**In this chapter, you will learn the following:**

✔ **How to identify if the use of Newton's Second Law is appropriate.**

✔ **How to apply Newton's Second Law to solve problems.**

✔ **How to test final derived relationships against expected behavior of the system.**

## Introduction

This chapter focuses on the conceptual understanding and application of one of the most famous equations in physics: Newton's Second Law, or $\sum \vec{F} = m\vec{a}$. Seeing and studying a systematic approach derived from the underlying principle will make this type of problem easy to solve and develop your intuition for how Newton's Second Law predicts behaviors of systems.

### Skills and Knowledge Needed to Solve Newton's Second Law Problems

◊ *Definition of a vector*: How a vector represents both magnitude and direction information.

◊ *Simple vector manipulation*: How to do simple vector addition and subtraction and perform scalar multiplication of a vector quantity.

◊ *Vector components*: How to express a vector in component form in a Cartesian coordinate system.

◊ *Word-problem skills*:

  • How to re-express physical and mathematical concepts expressed in words as mathematical expressions.

  • How to solve a system of equations symbolically.

◊ *Conceptual understanding*: How to frame expectations or make predictions of what will happen based on Newton's Second Law and Newton's Third Law.

## The Nature of Newton's Second Law Problems

Newton's Second Law problems are characterized by the relating of the acceleration of an object at some instant to the sum of the forces acting on that object at that instant. As such, Newton's Second Law is extremely useful for situations where the forces are constant during the time of interest. The other key aspect to these problems is that forces and acceleration are vectors, and thus keeping track of both magnitude and direction information is required.

Newton's Second Law is summarized mathematically and conceptually as follows:

**Newton's Second Law**

$$\sum \vec{F} = m\vec{a} \quad \text{or} \quad \sum \vec{F}/m = \vec{a}$$

*The vector sum of all of the forces acting on an object, $\sum \vec{F}$, divided by the amount of inertia that object has, m, has the same value (vectorially) as the acceleration, $\vec{a}$, the object experiences due to those forces.*

# Caveats and Subtleties of Newton's Second Law Problems

Newton's Second Law, $\sum \vec{F} = m\vec{a}$, is beautiful in how simply it can be expressed:

"*The sum of the forces acting on a body equals the product of the mass of the body and its acceleration.*"

However, as is often the case, the more succinct a physics concept or law can be expressed, the more powerful it is, and the greater the understanding of it is required to use it correctly. Newton's Second Law is one of these laws.

Fortunately, it is not difficult to remember all that is necessary to have success in using Newton's Second Law. It is when steps are skipped, both mathematical and conceptual, that errors creep in. As such, the best way to prevent these errors is to start always with the basics.

The three main aspects of Newton's Second Law to remember are:

- Newton's Second Law is a vector relation with both magnitude and direction information.

- Newton's Second Law is applied to one object at a time.

- Newton's Third Law is to be used after Newton's Second Law when a relation is needed between the magnitudes of the forces associated with an interaction between two objects.

## Vector Nature of Newton's Second Law

For most first-year physics students, vectors are quite new, and there is a temptation to use vector equations in the same manner as the scalar (non-vector) equations with which they are more familiar. For the simplest of Newton's Second Law problems, this will work. However, for problems even a bit more complicated, such an approach can fail. Therefore, understanding how to use Newton's Second Law starting in its vector form is needed and makes these problems much easier to do.

Unfortunately, the examples in most physics textbooks skip the steps in which the vector nature of Newton's Second Law is applied. Instead, these example solutions jump to the scalar equations that are then solved. However, it is in these skipped steps where the equations are created and the physics is applied. By skipping these steps, students are often left on their own to infer how to fill in these steps and inevitably make mistakes.

Fortunately, the remedy for this is easy: Always start with Newton's Second Law in its fundamental vector form. Doing so might add a few lines to the solution, but doing so is quick and will ensure errors are eliminated. The problem-solving steps detailed later include instructions for doing this.

To apply Newton's Second Law correctly, a free-body diagram of the forces is used first to capture the vector information correctly, both the magnitudes and directions. This is represented in the figure below.

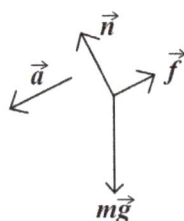

Here there are three forces acting on the object: $\vec{n}$, $\vec{f}$, and $m\vec{g}$. The acceleration vector $\vec{a}$ is also shown alongside the free-body diagram as a reminder of its orientation.

On the left side of the equation for Newton's Second Law, the vectors for the forces are summed. On the right is the product of mass and acceleration.

$$\sum \vec{F} = m\vec{a}$$
$$(\vec{n} + \vec{f} + m\vec{g}) = m\vec{a}$$

The next step is transforming this vector equation into scalar ones that can then be used with standard algebra. This is done by resolving each vector into component form using Cartesian coordinates, thereby providing an easy way to extract the equations for each direction of the coordinate system. Here, because the vectors are in a plane, only the $x$ and $y$ directions are needed.

The choice of coordinate system is arbitrary, but a good tactic is to pick an orientation for it so that as many of the vectors as possible are parallel to the axes of the coordinate system. Here that would mean the $x$-axis would be in the direction of $\vec{f}$, and the $y$-axis would be in the direction of $\vec{n}$. This way, only one vector, $m\vec{g}$, will have more than one non-zero component. Negative signs are introduced in this step, as necessary, to represent the direction of the vectors.

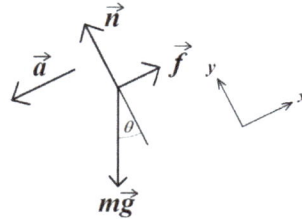

The vector equation then becomes:

$$\sum \vec{F} = m\vec{a}$$

$$\vec{n} + \vec{f} + m\vec{g} = m\vec{a}$$

$$(+n\,\hat{j}) + (+f\,\hat{i}) + (-mg\sin\theta\,\hat{i} - mg\cos\theta\,\hat{j}) = m(-a\hat{i})$$

Next, the last line is rearranged by collecting terms by unit vector so that it is easier to extract the equations for Newton's Second Law in the $x$ and $y$ directions.

$$\sum \vec{F} = m\vec{a}$$

$$\vec{n} + \vec{f} + m\vec{g} = m\vec{a}$$

$$n\hat{j} + f\hat{i} - mg\sin\theta\,\hat{i} - mg\cos\theta\,\hat{j} = -ma\hat{i}$$

$$(f - mg\sin\theta)\hat{i} + (n - mg\cos\theta)\hat{j} = (-ma)\hat{i} + (0)\hat{j}$$

$$x)\quad f - mg\sin\theta = -ma$$

$$y)\quad n - mg\cos\theta = 0$$

By following this approach, any ambiguity regarding the proper directions for the vectors (signs in the scalar equations) is eliminated so that the equations for the motion in each of the coordinate directions will be correct. (e.g., see the acceleration vector above.)

### Apply Newton's Second Law to One Object at a Time

Another common challenge to students is to determine which forces to include in their expression of Newton's Second Law. The way to overcome this is to remember always that when applying Newton's Second Law, only those forces directly acting on an object are included.

A good example is the situation in which there are two objects interacting through a string, such as in an Atwood machine.

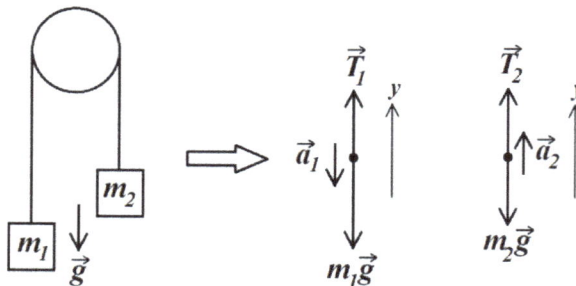

For the case above we have two objects on which two forces act on each, one due to tension of the string and one due to gravity. A common assumption is that the net force on object 1 should include the weight of object 2. This is understandable but incorrect. The best approach is to address in a separate equation that the magnitudes of the tension force on each object are the same. This is illustrated below with the equations of Newton's Second Law for each object being supplemented by additional equations.

$$\underline{m_1}$$ $$\underline{m_2}$$ because same string:

$$\sum \vec{F} = m\vec{a} \qquad \sum \vec{F} = m\vec{a} \qquad \overline{|\vec{a}_1| = |\vec{a}_2| \Rightarrow a_1 = a_2}$$

$$\vec{T}_1 + m_1\vec{g} = m_1\vec{a}_1 \qquad \vec{T}_2 + m_2\vec{g} = m_2\vec{a}_2 \qquad |\vec{T}_1| = |\vec{T}_2| \Rightarrow T_1 = T_2$$

$$+T_1 - m_1 g = -m_1 a \qquad +T_2 - m_2 g = +m_2 a$$

Note how only the forces acting on each of the objects is included in the corresponding expression for Newton's Second Law. Also note how the correct directions for each of the vectors and thus the signs in the scalar equations are easily identified from this set-up. Finally, the connections between the magnitudes of the tension forces and the accelerations are explicitly noted on the right. Starting with the vector form of Newton's Second Law, only including the actual forces acting on an object, and then writing any other supplemental information separately makes it easy to ensure the correct equations are created.

## Newton's Third Law

The need to apply Newton's Third law in Newton's Second Law problems is not uncommon. As such it is important to review how to use it in the context of problem-solving. Unfortunately, the classic description of Newton's Third Law is not very useful for this: "*For every action there is an equal and opposite reaction.*"

Instead, an alternate version of Newton's Third Law can be more informative: "*For every single interaction between two objects, each feels an equal and opposite force due to that one interaction.*" Mathematically this is represented by $\vec{F}_{12} = -\vec{F}_{21}$, which is read as "*The force on object 1 by object 2 has the same magnitude but opposite direction of the force on object 2 by object 1.*" With this definition, we then can write $|\vec{F}_{12}| = |\vec{F}_{21}|$ and thus $F_{12} = F_{21}$. Visually, this is represented below:

Situations where Newton's Third Law will need to be invoked is when the magnitudes of forces of interaction between two bodies must be related. Examples include when strings are attached between bodies, when two bodies are touching (pushing on each other), or even when the interaction is over a distance, such as the forces on the Earth and the Moon due to their mutual gravitational attraction. The case for when two objects are touching is illustrated below.

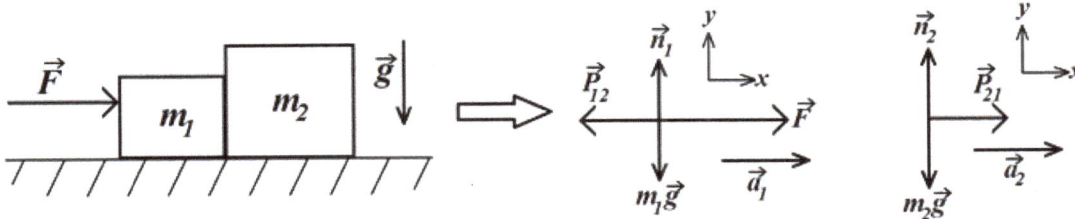

Here the contact forces are related by $\vec{P}_{12} = -\vec{P}_{21}$ and thus $P_{12} = P_{21}$ where also because the two blocks are touching and thus move together, $a_1 = a_2$.

The important point is that in the expression for Newton's Second Law for an object, only those forces that directly act on the object are included and then, as necessary, Newton's Third Law is invoked separately to relate the values of the magnitudes of forces acting on members of an interaction pair. By doing so, errors, especially sign errors, will be avoided.

## The Problem-Solving Steps for Newton's Second Law Problems

Below is a description of the basic steps for solving Newton's Second Law problems, written out so that you can understand the approach.

## Identify the Type of Problem

The most obvious step is to identify if the problem is one for which using Newton's Second Law is appropriate.

## Hallmarks of a Newton's Second Law Problem

- The statement of the problem references forces applied to one or more bodies and the acceleration (or lack of acceleration) of those bodies.

- The problem requests information about the instantaneous motion of an object or objects due to interactions with other bodies.

The first bullet point is pretty obvious but the second is perhaps less so. Whereas the conservation laws of physics can be used to characterize the *change* in the properties of motions or interactions (i.e., initial versus final speeds or states), Newton's Second Law tells us what is happening at some particular *instant in time* given the values of the forces applied, the accelerations, and the mass (amount of inertia) of the object being acted upon by those forces.

## Recall the Physics Principle: Newton's Second Law

- *Mathematical Form*: $\sum \vec{F} = m\vec{a}$ or $\sum \vec{F}/m = \vec{a}$

- *Conceptual Form*: "The vector sum of all of the forces interacting on an object, $\sum \vec{F}$, divided by the amount of inertia that object has, $m$, has the same value (vectorially) as the acceleration, $\vec{a}$, that object experiences due to those forces."

## The Free-Body Diagram; Applying Newton's Second Law Pictorially to a Problem

Because we are dealing with a vector problem, we must represent both magnitude and direction for our vector quantities. Drawing a free-body diagram accomplishes this along with helping to identify for later the most convenient coordinate system into which to decompose the vectors so that the algebra is easier to solve.

## Writing out the Vector Relationship

For each object separately, the forces acting on it are then included in the sum $\sum \vec{F}$, and this is set equal to the mass of the object multiplied by its acceleration: $\sum \vec{F} = m\vec{a}$. This is what Newton's Second Law is "telling" us to do: add up all the forces and set that equal to $m\vec{a}$ for each object.

*This step and the next one are not shown in most textbook example solutions even though they are the most important steps for solving a Newton's Second Law problem.*

## Creating the Scalar Version of Newton's Second Law

With the proper vector equation, the vectors in $\sum \vec{F} = m\vec{a}$ are transformed into unit vector form, followed by extracting the scalar relationships for each axis:

$$\vec{F}_1 \quad + \quad \vec{F}_2 \quad = m\vec{a} \qquad \text{Original vector equation}$$

$$\left(F_{1,x}\hat{i} + F_{1,y}\hat{j}\right) + \left(F_{2,x}\hat{i} + F_{2,y}\hat{j}\right) = m\left(a_x\hat{i} + a_y\hat{j}\right) \qquad \text{Rewritten equation using unit vectors}$$

$$\underbrace{\left(F_{1,x} + F_{2,x}\right)}_{x}\hat{i} + \underbrace{\left(F_{1,y} + F_{2,y}\right)}_{y}\hat{j} = \underbrace{\left(ma_x\right)}_{x}\hat{i} + \left(ma_y\right)\hat{j} \qquad \text{Terms collected by unit vector}$$

$$\boxed{F_{1,x} + F_{2,x} = ma_x} \quad \text{and} \quad \boxed{F_{1,y} + F_{2,y} = ma_y} \qquad \text{Final scalar equations}$$

This final step provides the scalar equations that are then much easier to use with our algebra tools. It is tempting to jump from the first line with the vectors to the final line with the scalar equations and insert the numerical values for each variable, but this will only lead to mistakes being made. Writing out the intermediate steps ensure that you will always get the correct scalar equations at the end while doing these steps symbolically makes it easier to check your own work.

## Doing the Algebra

The scalar equations are then solved simultaneously and symbolically using algebra for the parameter/variable of interest. Part of this step is to do a preliminary check that the number of equations is equal to the number of unknowns. If there are fewer equations than unknowns, then we would go back to see if there are any logical relationships or statements in the problem that we overlooked that can be expressed as equations.

## Re-Conceptualizing the Physics

This final step has several parts. They all contribute to making sure your work is correct and will help to maximize the grade you will get on what you turn in.

- Check Units: An easy and quick first step is to check if the units of an expression are what you expect them to be. If they are not correct, then the algebra should be checked. If correct, then you can be confident you are on the right track.

- Check Equation Predictions: Limiting-case values are inserted into the algebraic symbolic expression to see if the behavior is what would be expected for the special and often simpler case being examined. For example, if a ramp becomes horizontal, the acceleration of a block on it will go to zero. If not, then there is an error to be corrected.

- Reasonableness: When the numerical values for the parameters are inserted into the final expression, the reasonableness of the answer can be checked. For example, if a final speed is calculated to be in excess of the speed of light (i.e., if $v > 3.0 \times 10^8$ m/s), then you can be sure a mistake was made or the situation described is unphysical.

## Comment on Doing All Work with Symbolic Algebra

The example solutions will illustrate why it is highly recommended to do all of your problems symbolically instead of inserting numerical values early in the problem-solving process. The reasons for doing so are the following:

- It is much easier to check and correct your work if the algebra has been done symbolically.

- You may discover that some of the data that you thought might be necessary is not. For example, in friction problems, sometimes the mass is not needed.

- Your work will more likely be correct and thus earn full credit, and if there is an error, with more work shown, it will be easier for your instructors to justify awarding more partial credit.

# Newton's Second Law Problem-Solving Steps

<u>Hallmarks</u>: Forces applied to one or more bodies and the acceleration (or lack of acceleration) of those bodies mentioned.

**Newton's Second Law**

$$\sum \vec{F} = m\vec{a}$$

## Steps

1. **Identify components of system***

   - *Abstract the word problem* into mathematical expressions with a figure

   - *Apply the physics principle* by drawing a free-body diagram for each object upon which forces act (to aid in writing out full vector equation starting with $\sum \vec{F} = m\vec{a}$)*

   - *Define a coordinate system* for each object that makes defining the vectors simple

2. **Adapt and Apply Newton's Second Law to the problem***

   - *Write out the vector form of the equation* for Newton's Second Law for each object, using vector variables

   - *Rewrite vector equation in component form* using unit vectors for each of the vectors in the $\sum \vec{F} = m\vec{a}$ equation for each object

   - *Collect the terms* of vector equation(s) by unit vector

   - *Separate out the scalar equations* for each axis for each body's $\sum \vec{F} = m\vec{a}$ equation

3. **Identify any other relationships needed**

   - *Check* if you have the same number of equations as unknown variables

   - If necessary, *identify more data and/or equations* so that the number of equations equals the number of unknowns and so that system can be solved

   - *Write out the final set* of equations

4. **Do algebra to solve equations for parameter of interest**

   - *Use only variables*, not numerical values

5. **Check "physicalness" of final relationship**

   - *Perform unit analysis*

   - *Check predictions* of limiting behavior

6. **Evaluate expression using data** (if provided)

   - *Check physicalness and reasonableness* of the answer

\* *These are the key steps for this type of problem.*

# EXAMPLE 2.1

A ball of mass 5.0 kg hangs from a massless string connected to a block of mass 3.0 kg on a frictionless ramp inclined at 35° via a string passed over a massless pulley. What is the acceleration of the block along the ramp?

| SOLUTION: | ANNOTATION/COMMENT (STEP #): |
|---|---|
| | With forces and acceleration mentioned or inferred, we know Newton's Second Law is appropriate. |

data:

$m_1 = 5.0$ kg, $m_2 = 3.0$ kg, $\theta = 35°$,

$g = 9.8$ m/s², $a_2 = ?$

**Identify the raw data of the problem and the key request of the problem statement (1)**

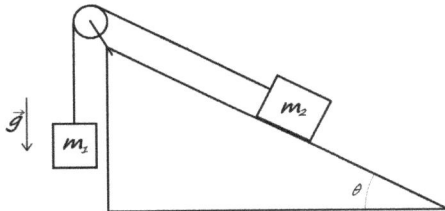

**Draw diagram to help identify all important parts (1)**

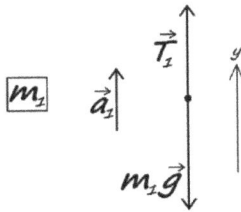

**Draw a Free Body Diagram and acceleration vector for each object and pick appropriate coordinate system for each (1)**

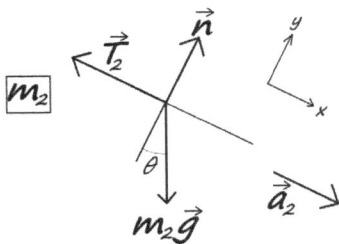

Note: These coordinate systems do not have to be the same for each object.

## SOLUTION:

**ANNOTATION/COMMENT (STEP #):**

**Translate vector equation $\sum \vec{F} = m\vec{a}$ for each object into scalar form (2)**

for $m_1$

$$\sum \vec{F} = m\vec{a}$$
$$\vec{T}_1 + m_1\vec{g} = m_1\vec{a}_1$$
$$(+T_1\hat{j}) + (-m_1 g\,\hat{j}) = m_1(+a_1\hat{j})$$

For $m_1$, the coordinate system is easy; up and down.

Note: The negative signs only start showing up in this step with the vectors translated into unit vector representation.

$$\boxed{+T_1 - m_1 g = +m_1 a_1}$$

Final equation for $m_1$

for $m_2$

$$\sum \vec{F} = m\vec{a}$$
$$\vec{T}_2 + \vec{n} + m_2\vec{g} = m_2\vec{a}_2$$

$$(-T_2\hat{i} + O\hat{j})$$
$$+(O\hat{i} + n\hat{j})$$
$$+(m_2 g \sin\theta\,\hat{i} - m_2 g \cos\theta)\hat{j}$$
$$= m_2(a_2\hat{i} + O\hat{j})$$

$$\hat{i}(-T_2 + m_2 g \sin\theta)$$
$$+\hat{j}(n - m_2 g \cos\theta)$$
$$= \hat{i}(m_2 a_2) + \hat{j}(O)$$

For $m_2$ on the incline, the coordinate system is more complicated requiring sine and cosine functions to write out the components of each vector.

Each vector is written out in component form, including negative signs. The zero components were included for completeness but are most often left out.

This trigonometry step is often confusing at first, but by using more, simpler steps, makes it easier.

The terms are collected by component direction.

$$\boxed{\begin{array}{ll} x) & -T_2 + m_2 g \sin\theta = m_2 a_2 \\ y) & n - m_2 g \cos\theta = O \end{array}}$$

Final equations for $m_2$

The use of "x)" and "y)" are used to help keep track from where each equation originated.

## SOLUTION:

| variable | known? |
|----------|--------|
| $T_1$ | no |
| $T_2$ | no |
| $m_1$ | yes |
| $m_2$ | yes |
| $\theta$ | yes |
| $g$ | yes |
| $n$ | no |
| $a_1$ | no |
| $a_2$ | no |

Unknowns = 5, Equations = 3

Same string:

$$\therefore \left| \vec{a}_1 \right| = \left| \vec{a}_2 \right| \rightarrow a_1 = a_2$$

$$\text{let } a_2 = a$$

$$\therefore \left| \vec{T}_1 \right| = \left| \vec{T}_2 \right| \rightarrow T_1 = T_2$$

$$\text{let } T_2 = T$$

$$\boxed{\begin{aligned} T - m_1 g &= m_1 a \\ -T + m_2 g \sin\theta &= m_2 a \\ n - m_2 g \cos\theta &= 0 \end{aligned}}$$

Unknowns: $T, a, n$

## ANNOTATION/COMMENT (STEP #):

### Check if number of scalar equations equals number of unknowns (3)

Note: Very often, such a table and the listing of the number of unknowns and equations is not shown and instead is done in one's head. However, for more complicated problems, writing out such a table can be very useful.

With 5 unknowns and 3 equations, we need at least 2 more equations or more data.

Because the masses are connected by the same string, we can say that *the magnitude of the accelerations must be the same*. Similarly, because the pulley is massless (no mass to accelerate), *the magnitude of the force exerted on each block by the string must be the same*. This gives us the extra equations that we need (4 of them in total, including the introduction of the new variables $a$ and $T$).

Note: If the direction $\vec{a}_1$ had been instead chosen so that it pointed downward, we would have written $a_1 = -a_2$, and the algebra would have still worked.

Final set of equations

Here, the substitutions $T_1 = T_2 = T$ and $a_1 = a_2 = a$ were executed. Now we have 3 equations and 3 unknowns.

**SOLUTION:**

$$T - m_1 g = m_1 a$$
$$-T + m_2 g \sin\theta = m_2 a$$
$$n - m_2 g \cos\theta = 0$$

$$T = m_1 g + m_1 a$$

$$-(m_1 g + m_1 a) + m_2 g \sin\theta = m_2 a$$

$$-m_1 g - m_1 a + m_2 g \sin\theta = m_2 a$$
$$-m_1 g + m_2 g \sin\theta = m_2 a + m_1 a$$
$$g(m_2 \sin\theta - m_1) = (m_2 + m_1) a$$

$$a = g \frac{(m_2 \sin\theta - m_1)}{(m_2 + m_1)}$$

$$[a] = \left[ g \frac{(m_2 \sin\theta - m_1)}{(m_2 + m_1)} \right]$$
$$= [g] \frac{[(m_2 \sin\theta - m_1)]}{[(m_2 + m_1)]}$$
$$= \left(\frac{m}{s^2}\right) \frac{(kg \cdot (1) + kg)}{(kg + kg)} = \left(\frac{m}{s^2}\right) \left|\frac{kg}{kg}\right|$$

$$[a] = \frac{m}{s^2} \quad \checkmark$$

**ANNOTATION/COMMENT (STEP #):**

**Use algebra to solve the system of equations simultaneously for parameter of interest (4)**

Because $n$, the magnitude of the normal force, does not show up in the equations that include $a$ or $T$, we can expect that we likely will not need to use $n$ to find the acceleration.

First equation is solved for $T$ and then that result is substituted into the second equation. The substitution is shown with the parentheses.

The resulting equation is now solved for the variable of interest: $a$.

Final expression for the acceleration

Note: We could have just as easily eliminated $a$ and solved for $T$ if that was what was requested in the problem.

**Check units (5)**
Checking if the units of $a$ are m/s$^2$.

Note: Checking units is normally not shown in assignments that are turned in. It is shown here to illustrate how it is done. It is important to do unit analysis whether it is to be shown or not, as it gives a good way to check your work and catch your own errors.

The units are as expected. Therefore, it is worth moving to the next step.

**SOLUTION:**

$$m_2 \to 0:$$

$$a = g\frac{\left((0)\sin\theta - m_1\right)}{\left((0) + m_1\right)}$$

$$a = g\frac{\left(-\cancel{m_1}\right)}{\left(\cancel{m_1}\right)} = -g$$

$$a = -g \checkmark$$

$$m_1 \to 0:$$

$$a = g\frac{(m_2\sin\theta - (0))}{(m_2 + (0))}$$

$$a = g\frac{\cancel{m_2}\sin\theta}{\cancel{m_2}} = g\sin\theta$$

$$a = g\sin\theta \checkmark$$

$$\theta \to 90°:$$

$$a = g\frac{(m_2\sin(90°) - m_1)}{(m_2 + m_1)}$$

$$a = g\frac{(m_2(1) - m_1)}{(m_2 + m_1)}$$

$$a = g\frac{(m_2 - m_1)}{(m_2 + m_1)}, \quad \text{and}$$

$$\text{sign}(a) = \text{sign}(m_2 - m_1) \checkmark$$

**ANNOTATION/COMMENT (STEP #):**

**Check physicalness (5)**

Note: Checking predictions is often done in one's head and is not shown on work that is turned in. But it should still be done as a useful check.

Recall that $a_1$ was defined as positive in the upwards direction for $m_1$, so its sign is correct for downward free-fall.

With only $m_1$ connected, expect $m_1$ to free fall, i.e., $a$ becomes $a = -g$.

Here we would expect to see the equation predict the motion of a single block sliding down a ramp.

With only $m_2$ connected, $m_2$ slides without friction down the ramp, in the 'positive' direction, i.e., $a$ becomes $a = +g\sin\theta$.

$\theta = 90°$ corresponds to a vertical ramp with only the pulley supporting the masses:

If $m_2 > m_1$, then the acceleration would be positive, consistent with how the direction for $a_2$ was defined as downward for $m_2$.

Similarly, if $m_2 < m_1$, then the acceleration would be negative because the masses would move in the opposite direction than originally defined.

**SOLUTION:**

**ANNOTATION/COMMENT (STEP #):**

**Evaluate final expression numerically (6)**

$$a = g \frac{(m_2 \sin\theta - m_1)}{(m_2 + m_1)}$$

$$a = \left(9.8 \frac{m}{s^2}\right) \frac{(3.0 \text{ kg} \sin 35° - 5.0 \text{ kg})}{(3.0 \text{ kg} + 5.0 \text{ kg})}$$

$$= \left(9.8 \frac{m}{s^2}\right) \frac{((3.0 \text{ kg})(0.57) - 5.0 \text{ kg})}{(3.0 \text{ kg} + 5.0 \text{ kg})}$$

$$= \left(9.8 \frac{m}{s^2}\right)\left(\frac{-3.3 \cancel{kg}}{8.0 \cancel{kg}}\right)$$

Note: The cancellation and collecting of the units as part of the numerical evaluation acts as a final double-check of the units.

$$= \left(9.8 \frac{m}{s^2}\right)(-0.413)$$

$$\boxed{a = -4.0 \frac{m}{s^2}}$$

Final numerical answer

This answer makes sense because its magnitude is between zero and $g$ and is also negative, corresponding to the mass on the incline moving 'uphill' due to the other mass being larger and that the uphill direction having been defined as the negative direction.

$$T = ?$$

For "fun," we can now also solve for the magnitude of the tension.

$$T = m_1 g + m_1(a)$$

$$T = m_1 g + m_1 g \left(\frac{(m_2 \sin\theta - m_1)}{(m_2 + m_1)}\right)$$

$$T = m_1 g \left(1 + \frac{(m_2 \sin\theta - m_1)}{(m_2 + m_1)}\right)$$

Here, in the equation solved for $T$, a substitution is made for $a$ using the second equation, and then the first equation can be solved for $T$.

$$T = (5.0 \text{ kg})(9.8 \, m/s^2)$$
$$\times \left(1 + \frac{(3.0 \cancel{kg}) \sin 35° - (5.0 \cancel{kg})}{(3.0 \cancel{kg} + (5.0 \cancel{kg}))}\right)$$

If only the numerical value for $a$ was desired, we could have substituted it directly to calculate $T$.

$$T = \left(49 \frac{kg \cdot m}{s^2}\right)\left(1 + \frac{-3.3}{8.0}\right)\frac{N}{\cancel{kg \cdot m/s^2}}$$

$$\boxed{T = 29 \text{ N}}$$

To be physical, this value can only be positive. In other words: "You can't push on a rope."

# EXAMPLE 2.2

A conical pendulum consists of a string attached to the top of a vertical pole at the end of which is a mass that is traveling in a circle so that the string sweeps out the surface of a cone. The mass of the object is 0.20 kg, the length of the string is 0.30 m, and its speed is 0.86 m/sec. What is the tension in the string? What is the angle the string makes with the pole?

| SOLUTION: | ANNOTATION/COMMENT (STEP #): |
|---|---|
| | Forces mentioned or inferred (gravity, tension) and acceleration (uniform circular motion); Newton's Second Law is appropriate |

*data:*

$m = 0.20$ kg, $\ell = 0.30$ m

$g = 9.8$ m/s², $v = 0.86$ m/s,

$T = ?$, $\theta = ?$

Identify the raw data of the problem and the key request (1)

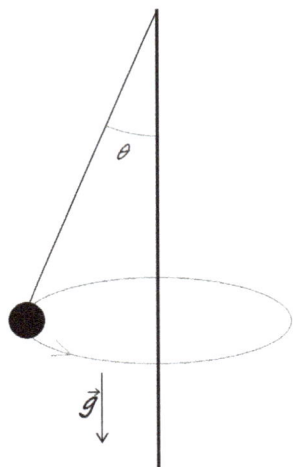

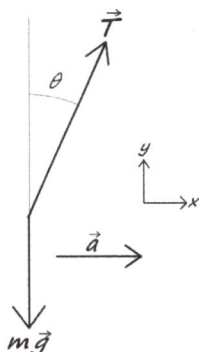

Draw diagram to help identify all important parts (1)

Draw free-body diagram to identify the forces acting on the body and include the expected acceleration vector (1)

Choose a coordinate system expected to be convenient for doing the later algebra (1)

Note: Although the directions for the tension force and the acceleration are always changing, at all times their relative directions are fixed.

Note: For circular motion, there is always a component of the acceleration that is inward. When it is a *uniform* circular motion, the acceleration is only inward.

**SOLUTION:**

$$\sum \vec{F} = m\vec{a}$$

$$\vec{T} + m\vec{g} = m\vec{a}$$

$$\left(T \sin\theta \,\hat{\imath} + T \cos\theta \,\hat{\jmath}\right) +$$

$$\left(-mg\,\hat{\jmath}\right) = m\left(a\hat{\imath} + O\hat{\jmath}\right)$$

$$\hat{\imath}\left(T\sin\theta\right) + \hat{\jmath}\left(T\cos\theta - mg\right)$$

$$= \hat{\imath}\left(ma\right) + \hat{\jmath}\left(O\right)$$

x) $T \sin\theta = ma$

y) $T \cos\theta - mg = 0$

| variable | known? |
|----------|--------|
| $T$ | no |
| $\theta$ | no |
| $\ell$ | yes |
| $a$ | no |
| $m$ | yes |
| $g$ | yes |
| $v$ | yes |

Equations = 2, unknowns = 3

<u>UCM:</u> $a = \dfrac{v^2}{r}$

$r = \ell \sin\theta$

**ANNOTATION/COMMENT (STEP #):**

**Rewrite $\sum \vec{F} = m\vec{a}$ using information from the free-body diagram and then transform vectors into unit vector form so that the scalar equations can be easily identified (2)**

Because the angle is defined by the orientation of the string with vertical instead of horizontal, we have $T\sin\theta$ for the x-component and $T\cos\theta$ for the y-component.

The scalar equations for each axis

**Check if number of scalar equations equals number of unknowns (3)**

Note: Writing out a table of knowns and unknowns is not normally necessary and can instead be done by inspection or "in one's head," but it still should be done.

We have 2 equations and 3 unknowns, so at least one more equation is needed.

To use Newton's second law, we also need a way to find $a$. Use relationship for acceleration for uniform circular motion. But this introduces another variable, $r$.

The radius $r$ can be related to $\ell$ and $\theta$ by geometry.

## SOLUTION:

| variable | known? |
|----------|--------|
| $T$ | no |
| $\theta$ | no |
| $\ell$ | yes |
| $a$ | no |
| $m$ | yes |
| $g$ | yes |
| $v$ | yes |
| $r$ | no |

4 unknowns, 4 equations

$$T \sin \theta = ma$$
$$T \cos \theta - mg = 0$$
$$a = v^2/r, \quad r = \ell \sin \theta$$

$$T \sin \theta = a$$
$$T \sin \theta = m \left( \frac{v^2}{r} \right)$$
$$T \sin \theta = \frac{mv^2}{\left( \ell \sin \theta \right)}$$

$$\boxed{T = \frac{mv^2}{\ell \sin^2 \theta}}$$

$$T \cos \theta = mg$$
$$\boxed{T = \frac{mg}{\cos \theta}}$$

## ANNOTATION/COMMENT (STEP #):

Updated table with number of equations and unknowns

With 4 equations and 4 unknowns, expect the equations to be solvable.

Final system of equations

**Solve system of equations for desired variables (4)**

Attempt to solve for the tension so to eliminate it and find the angle

Substituting the expression for acceleration and radius. (shown by using parentheses)

Solving for tension using the second equation.

**SOLUTION:**

**ANNOTATION/COMMENT (STEP #):**

Combining the expressions for the tension to find the angle

$$\left(\frac{mv^2}{\ell \sin^2 \theta}\right) = T = \left(\frac{mg}{\cos \theta}\right)$$

$$\frac{\cancel{m}v^2}{\ell \sin^2 \theta} = \frac{\cancel{m}g}{\cos \theta}$$

$$\frac{v^2}{\ell \sin^2 \theta} = \frac{g}{\cos \theta}$$

$$\frac{\cos \theta}{\sin^2 \theta} = \frac{g\ell}{v^2}$$

$$\boxed{\sin \theta \, \tan \theta = \frac{v^2}{g\ell}}$$

The result is a transcendental equation for $\theta$; i.e., it cannot be solved for $\theta$ using algebra alone.

$$\boxed{\cos^2 \theta + \sin^2 \theta = 1}$$

Starting over with new approach using the Pythagorean Theorem to take advantage of sine and cosine expressions.

$$\left.\begin{array}{l} T \cos \theta = mg \\ \cos \theta = \dfrac{mg}{T} \end{array}\right\} \rightarrow \left(\cos^2 \theta\right) = \left(\frac{mg}{T}\right)^2$$

Solve free-body diagram equations for $\sin^2\theta$ and $\cos^2\theta$.

$$\left.\begin{array}{l} T \sin \theta = m\left(\dfrac{v^2}{r}\right) \\ T \sin \theta = \left(\dfrac{mv^2}{\ell \sin \theta}\right) \end{array}\right\} \rightarrow \left(\sin^2 \theta\right) = \frac{mv^2}{\ell T}$$

Note: Sometimes a new mathematical approach is necessary.

$$\left(\left[\frac{mg}{T}\right]^2\right) + \left(\frac{mv^2}{\ell T}\right) = 1$$

Substitute results into Pythagorean Theorem and solve in terms of $T$.

$$\frac{m^2 g^2}{T^2} + \frac{mv^2}{\ell T} = 1$$

$$m^2 g^2 + \frac{mv^2}{\ell} T = T^2$$

$$T^2 - \frac{mv^2}{\ell} T - m^2 g^2 = 0$$

$$\boxed{T^2 + \left(-\frac{mv^2}{\ell}\right) T + \left(-m^2 g^2\right) = 0}$$

In the final line, have a polynomial that can be solved using the Quadratic Formula.

**SOLUTION:**

$$T^2 + \left(-\frac{mv^2}{\ell}\right)T + \left(m^2g^2\right) = 0$$

where

$$a = 1, \quad b = \left(-\frac{mv^2}{\ell}\right), \quad c = \left(-m^2g^2\right)$$

and

$$T = \frac{1}{2a}\left(-b \pm \sqrt{b^2 - 4ac}\right)$$

$$T = \frac{1}{2(1)}\left(\left(\frac{mv^2}{\ell}\right) \pm \sqrt{\left(\frac{mv^2}{\ell}\right)^2 - 4(1)\left(-m^2g^2\right)}\right)$$

$$= \frac{1}{2}\left(\left(\frac{mv^2}{\ell}\right) \pm \sqrt{\left(\frac{mv^2}{\ell}\right)^2 + 4m^2g^2}\right)$$

$$= \left(\frac{mv^2}{2\ell}\right)\left(1 \pm \sqrt{1 + 4m^2g^2\left(\frac{\ell^2}{m^2v^4}\right)}\right)$$

$$T = \left(\frac{mv^2}{2\ell}\right)\left(1 \pm \sqrt{1 + \left(\frac{4g^2\ell^2}{v^4}\right)}\right)$$

**ANNOTATION/COMMENT (STEP #):**

Identifying coefficients for Quadratic Formula

Substituting coefficients into Quadratic Formula

Final Expression for the tension

**Check units (5)**

Because the equation is complicated will do check in two parts: the prefactor first.

Because the second factor appears unitless, we expect the prefactor to have units of Newtons.

> Note: Checking units is not normally shown on the work that is turned in. But it should always be done to check your work.

Units of prefactor are Newtons as expected.

$$\left[\frac{mv^2}{2\ell}\right] = \frac{[m][v]^2}{[2][\ell]}$$

$$= \frac{[kg][m^2/s^2]}{(1)[m]}$$

$$= kg\frac{m}{s^2}\left(N/kg\frac{m}{s^2}\right)$$

$$\left[\frac{mv^2}{2\ell}\right] = N \checkmark$$

| SOLUTION: | ANNOTATION/COMMENT (STEP #): |
|---|---|

$$\left[\frac{4g^2\ell^2}{v^4}\right] = \frac{[4][g^2][\ell^2]}{[v^4]}$$

Second factor; The argument of the square root function must be unitless because there is a numeral in the argument. Therefore, the algebraic term in the argument of the square root function must also be unitless.

$$= \frac{(1)(m^2/s^4)(m^2)}{(m^4/s^4)}$$

$$= \frac{m^4 s^4}{m^4 s^4} = 1$$

The argument of the square root term is unitless as expected.

$$\left[\frac{4g^2\ell^2}{v^4}\right] = 1 \checkmark$$

Therefore, the whole term has units of Newtons as expected.

Checking units for equation for the angle

$$T\cos\theta = mg$$

$$\boxed{\cos\theta = \frac{mg}{T}}$$

Trigonometric functions are always unitless (as are angles).

$$[\cos\theta] = \left[\frac{mg}{T}\right] = \frac{[m][g]}{[T]}$$

$$= \frac{kg(m/s^2)}{N} = 1$$

$$[\cos\theta] = 1 \checkmark$$

Units for angle equation are as expected.

**Check physicalness (5)**

For larger tension with a fixed mass, the expression predicts that $\cos\theta$ will get smaller, which for angles between 0° and 90° corresponds to larger angles.

$$\underline{T \to larger:}$$

$$\cos\theta = \frac{mg}{T} \to smaller \checkmark$$

Behavior is as expected because for $T \gg mg$, gravity is relatively negligible, and the mass swings in a horizontal circle.

Note: Checking physicalness is not normally shown and can often be done by inspection. But it still should be done as a way to check if the expression is likely correct.

| SOLUTION: | ANNOTATION/COMMENT (STEP #): |
|---|---|

**SOLUTION:**

$$T = \left(\frac{mv^2}{2\ell}\right)\left[1 \pm \sqrt{1 + \left(\frac{4g^2\ell^2}{v^4}\right)}\right]$$

'+' or '–' sign?
because

$$\sqrt{1 + \left(\frac{4g^2\ell^2}{v^4}\right)} \sim \left(1 + \delta\right) > 1,$$

then $1 + \sqrt{1 + \left(\frac{4g^2\ell^2}{v^4}\right)} > 2$

and $1 - \sqrt{1 + \left(\frac{4g^2\ell^2}{v^4}\right)} < 0$

∴ only '+' sign is physical

for $v \rightarrow$ larger:

$$T = \left(\frac{mv^2}{2\ell}\right)\left[1 + \sqrt{1 + \left(\frac{4g^2\ell^2}{v^4}\right)}\right]$$

$$\sim \left(\frac{mv^2}{2\ell}\right)\left[1 + \sqrt{1 + \left(\frac{4g^2\ell^2}{\infty}\right)}\right]$$

$$\sim \left(\frac{mv^2}{2\ell}\right)\left[1 + \sqrt{1 + (0)}\right]$$

$$\sim \left(\frac{mv^2}{2\ell}\right)(1 + 1) \sim m\frac{v^2}{\ell}$$

$T \rightarrow$ larger $\rightarrow m\,(v^2/\ell)$ ✓

**ANNOTATION/COMMENT (STEP #):**

Checking expression for tension for its predictions.

Two possible solutions. Are both or only one possible?

Because tensions can only be greater than zero, the negative sign answer is unphysical; i.e., "*You can't push on a rope.*"

Only the solution with the '+' sign is physical.

Check predictions of expression for $T$ due to variations in $v$ and $l$.

Because acceleration goes as $v^2$, if $v$ increases, then the force with a component in the direction of the acceleration must increase too.

Not only does the tension get larger ($T \sim v^2$), the expression for the tension takes on the form expected if gravity were negligible: $T = ma = m\,(v^2/l)$ where $r = l$, and the object rotates in a horizontal plane.

$T$ varies with $v$ as expected

| SOLUTION: | ANNOTATION/COMMENT (STEP #): |
|---|---|

for $v \to$ smaller:

$$T = \frac{1}{2}\left[\left(\frac{mv^2}{\ell}\right) + \sqrt{\left(\frac{mv^2}{\ell}\right)^2 + 4m^2 g^2}\right]$$

$$\sim \frac{1}{2}\left[\left(\frac{m(0)}{\ell}\right) + \sqrt{\left(\frac{m(0)}{\ell}\right)^2 + 4m^2 g^2}\right]$$

$$\sim \frac{1}{2}\left[(0) + \sqrt{(0) + 4m^2 g^2}\right]$$

$$\sim \frac{1}{2}\sqrt{4m^2 g^2} \sim \sqrt{(mg)^2}$$

$$\sim mg$$

$v \to$ smaller; $T \sim mg$ ✓

If the speed is reduced and approaches zero, we would expect the tension to get close to equaling the weight only; i.e., the mass is mostly just hanging straight down.

Change in $T$ as a function of $v$ is as expected.

**Evaluate final expressions numerically (6)**

$$T = \left(\frac{mv^2}{2\ell}\right)\left(1 + \sqrt{1 + \left(\frac{4g^2\ell^2}{v^4}\right)}\right)$$

$$= \frac{(0.20\ \text{kg})(0.86\ \text{m/s})^2}{2(0.30\ \text{m})}$$

$$+ \left(1 + \sqrt{1 + \left(\frac{4(9.8\ \text{m/s})^2(0.30\ \text{m})^2}{(0.86\ \text{m/s})^4}\right)}\right)$$

$$= \left(0.247\ \text{kg}\,\frac{\text{m}}{\text{s}^2}\right)\left(1 + \sqrt{1 + 63.2}\right)$$

$$= 2.226\ \text{N}$$

$$\boxed{T = 2.2\ \text{N}}$$

Final value for tension

$$\cos\theta = \frac{mg}{T}$$

$$\theta = \cos^{-1}\left(\frac{mg}{T}\right)$$

$$= \cos^{-1}\left(\frac{(0.20\ \text{kg})(9.8\ \text{m/s}^2)}{2.226\ \text{N}}\right)$$

$$= \cos^{-1}(0.881) \to \theta = 28.30°$$

$$\boxed{\theta = 28°}$$

Calculating the angle using the value for the tension calculated above.

This is between 0 and 90°, so this is an appropriate value.

Final value for the angle

# EXAMPLE 2.3

Two blocks are on a frictionless surface, with one next to the other. One is pushed by a force, and they both accelerate. What is the magnitude of the force of contact of the first one on the second if the applied force is 3.0 N and

a) $m_1 = 0.10$ kg, $m_2 = 1.0$ kg
b) $m_1 = 1.0$ kg, $m_2 = 1.0$ kg
c) $m_1 = 1.0$ kg, $m_2 = 0.10$ kg

**SOLUTION:**

**ANNOTATION/COMMENT (STEP #):**

Forces and accelerations in system. Newton's Second Law is appropriate to use.

$P_2 = ?$, $F = 3.0$ N

Identify the raw data of the problem and the key request (1)

a) $m_1 = 0.10$ kg, $m_2 = 1.0$ kg
b) $m_1 = 1.0$ kg, $m_2 = 1.0$ kg
c) $m_1 = 1.0$ kg, $m_2 = 0.10$ kg

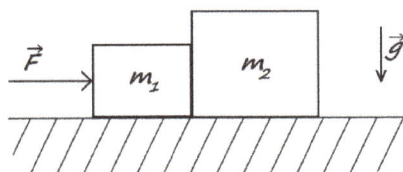

Draw a diagram to help identify all important parts (1)

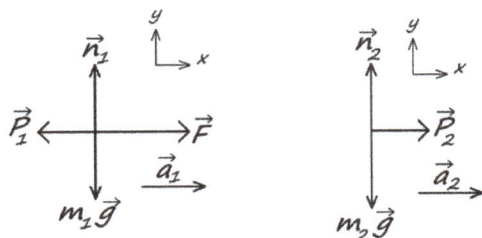

Draw a free-body diagram to identify the forces acting on the body and the expected acceleration vector (1)

Choose a coordinate system for each free-body diagram (1)

$m_1$:

$$\sum \vec{F} = m\vec{a}$$

$$\vec{P}_1 + \vec{n}_1 + \vec{F}_1 + m_1\vec{g} = m_1\vec{a}_1$$

$$-P_1\hat{i} + n_1\hat{j} + F\hat{i} - m_1g\hat{j} = m_1a_1\hat{i}$$

Rewrite $\sum \vec{F} = m\vec{a}$ using information from the free-body diagram and then apply chosen coordinate system so that the scalar equations can be easily identified (2)

**SOLUTION:**

**ANNOTATION/COMMENT (STEP #):**

$$\boxed{\begin{aligned} x) \quad &-P_1 + F = m_1 a_1 \\ y) \quad &n_1 - m_1 g = 0 \end{aligned}}$$

Final equations for $m_1$

$m_2$:

$$\sum \vec{F} = m\vec{a}$$

$$\vec{P_2} + \vec{n_2} + m_2 \vec{g} = m_2 \vec{a_2}$$

$$+P_2 \hat{i} + n_2 \hat{j} - m_2 g \hat{j} = m_2 a_2 \hat{i}$$

**Rewrite** $\sum \vec{F} = m\vec{a}$ **using information from the free-body diagram (2) and then apply chosen coordinate system so that the scalar equations can be easily identified (2)**

$$\boxed{\begin{aligned} x) \quad & P_2 = m_2 a_2 \\ y) \quad & n_2 - m_2 g = 0 \end{aligned}}$$

Final equations for $m_2$

**Check if number of equations equals number of unknowns (3)**

| variable | known? |
|----------|--------|
| $P_1$ | no |
| $P_2$ | no |
| $F$ | yes |
| $a_1$ | no |
| $a_2$ | no |
| $m_1$ | yes |
| $m_2$ | yes |
| $g$ | yes |
| $n_1$ | no |
| $n_2$ | no |

4 equations, 6 unknowns

Note: Checking if the number of equations is equal to the number of unknowns is normally not shown but should at least be done "in one's head." For more complicated problems, writing out a table of variables along with the equations can be very helpful.

Need 2 more equations

**Apply any other relationships or physics (3)**

N-3 and blocks touching:

$$\left| \vec{P_1} \right| = \left| \vec{P_2} \right| \quad \text{or} \quad P_1 = P_2 = P$$

$$\left| \vec{a_1} \right| = \left| \vec{a_2} \right| \quad \text{or} \quad a_1 = a_2 = a$$

Newton's Third Law and that the blocks are touching are used to create new equations. Also introduced new variables $a$ and $P$.

Additional required equations

## SOLUTION:

1) $-P + F = m_1 a$
2) $P = m_2 a$

*unneeded:*

$n_1 - m_1 g = 0, \; n_2 - m_2 g = 0$

$a = P/m_2$

$-P + F = m_1 (P/m_2)$

$F = \dfrac{m_1}{m_2} P + P$

$F = P(m_1/m_2 + 1)$

$P = \dfrac{F}{m_1/m_2 + 1}$

$[P] = \dfrac{[F]}{[m_1]/[m_2] + 1}$

$= \dfrac{N}{(kg)/(kg) + 1} = \dfrac{N}{[1+1]} = \dfrac{N}{1}$

$[P] = N$ ✓

## ANNOTATION/COMMENT (STEP #):

New simplified equations

Used new variables $P$ and $a$ that were introduced earlier to simplify the expressions further.

The equations for the normal forces are mathematically trivial, i.e., simple, and do not provide any useful information for this problem and are extra.

**Do algebra to find expression for desired variable (4)**

Find an expression for $a$ to substitute into the other equation.

Final relationship for $P$

Expressions for $n_1$, $n_2$, and $a$ not requested.

Note: We often have unknowns for which we are not asked to find expressions, although we could find an expression for each if we wanted.

**Check units (5)**

Checking if units for $P$ are in Newtons.

Note: Checking units is often not included in assignment solutions, but this step should always be done at least by inspection.

Units for $P$ are as expected.

| SOLUTION: | ANNOTATION/COMMENT (STEP #): |
|---|---|

**Check limiting cases (5)**

$\underline{m_1 >> m_2 \text{ or } (m_1/m_2) >> 1}$

$$P = \frac{F}{m_1/m_2 + 1} \sim \frac{F}{(\infty) + 1} \sim \frac{F}{\infty} \sim 0$$

$P \sim 0 \checkmark$

If $m_2 << m_1$, then the contact force does not need to be very large to accelerate the second block.

Note: Creating a unitless ratio of parameters can be very useful for checking limiting behavior.

$\underline{m_1 << m_2 \text{ or } (m_1/m_2) << 1}$

$$P = \frac{F}{m_1/m_2 + 1} \sim \frac{F}{(0) + 1} \sim \frac{F}{1} \sim F$$

$P \sim F \checkmark$

If $m_2 >> m_1$, then the contact force must be large to accelerate the second block.

Limiting case behavior is as expected.

Note: Checking limits are normally not included in assignment solutions, but it should always be done.

**Evaluate the expression using the data provided (6)**

a) $\underline{m_1 = 0.10 \text{ kg and } m_2 = 1.0 \text{ kg}}$

$$P = \frac{F}{m_1/m_2 + 1} = \frac{3.0 \text{ N}}{(0.10 \text{ kg}/1.0 \text{ kg}) + 1}$$

$$= \frac{3.0 \text{ N}}{(0.10 + 1)} = \boxed{2.7 \text{ N}}$$

The final values for each set of mass values.

The ratios for the contact force vary by a factor of 10, just as the ratio of the mass of the blocks vary, consistent with our expectations.

b) $\underline{m_1 = 1.0 \text{ kg and } m_2 = 1.0 \text{ kg}}$

$$P = \frac{F}{m_1/m_2 + 1} = \frac{3.0 \text{ N}}{(1.0 \text{ kg}/1.0 \text{ kg}) + 1}$$

$$= \frac{3.0 \text{ N}}{(1.0 + 1)} = \boxed{1.5 \text{ N}}$$

Note: Sometimes, like here, the final answer only depends on the ratio or relative magnitudes of values.

c) $\underline{m_1 = 1.0 \text{ kg and } m_2 = 0.10 \text{ kg}}$

$$P = \frac{F}{m_1/m_2 + 1} = \frac{3.0 \text{ N}}{(1.0 \text{ kg}/0.10 \text{ kg}) + 1}$$

$$= \frac{3.0 \text{ N}}{(10 + 1)} = \boxed{0.27 \text{ N}}$$

# EXAMPLE 2.4

45

A 5.0 kg box is pulled on a level surface with a force of 10 N directed upward from horizontal. What must the value of the friction coefficient between the box and floor be when the angle is 15° if the acceleration is 1.2 m/s²? What must this value be if the angle is 75°?

## SOLUTION:

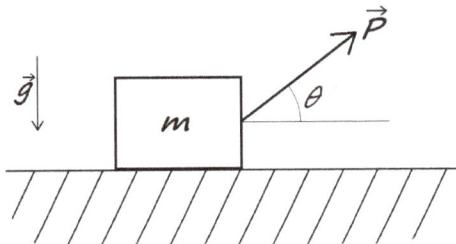

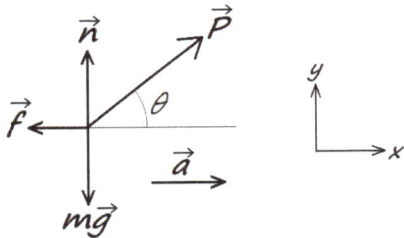

### ANNOTATION/COMMENT (STEP #):

Newton's Second Law is appropriate; forces (gravity, tension, friction) and acceleration mentioned or inferred

data:

$m = 5.0$ kg, $P = 10$ N,

$g = 9.8\dfrac{m}{s^2}$, $a = 1.2$ m/s²,

$\theta = 15°, 75°, \mu = ?$

Identify the raw data of the problem and the key request (1)

Draw diagram to help identify all important parts (1)

Draw a free-body diagram to identify the forces acting on the body and include the expected acceleration vector (1)

Choose a coordinate system expected to be convenient for doing the later algebra (1)

| SOLUTION: | ANNOTATION/COMMENT (STEP #): |

**SOLUTION:**

$$\sum \vec{F} = m\vec{a}$$
$$\vec{f} + \vec{n} + \vec{P} + m\vec{g} = m\vec{a}$$
$$(-f\,\hat{i}) + (n\,\hat{j})$$
$$+ (P\cos\theta\,\hat{i} + P\sin\theta\,\hat{j}) + (-mg\,\hat{j})$$
$$= m(a\,\hat{i} + 0\,\hat{j})$$
$$\hat{i}(-f + P\cos\theta) + \hat{j}(n + P\sin\theta - mg)$$
$$= \hat{i}(ma) + \hat{j}(0)$$

| | |
|---|---|
| x) | $-f + P\cos\theta = ma$ |
| y) | $n + P\sin\theta - mg = 0$ |

**ANNOTATION/COMMENT (STEP #):**

**Rewrite $\sum \vec{F} = m\vec{a}$ using information from the free-body diagram and then apply chosen coordinate system so that the scalar equations can be easily identified (2)**

The final equations for the block

**Check the number of equations and number of unknowns (3)**

| variable | known? |
|----------|--------|
| f | no |
| P | yes |
| n | no |
| m | yes |
| g | yes |
| θ | yes |
| a | no |

2 equations, 3 unknowns

Note: Checking if the number of equations is equal to the number of unknowns is normally not shown but should at least be done "in one's head." For more complicated problems, writing out a table of variables along with the equations can be very helpful.

Need one more equation.

$$f = \mu n$$

Extra equation to use

Note: Whenever friction is involved, $f = \mu n$ will likely be needed.

| |
|---|
| $-f + P\cos\theta = ma$ |
| $n + P\sin\theta - mg = 0$ |
| $f = \mu n$ |

Final set of equations to solve
Preliminary physics is done. Algebra now starts.

| **SOLUTION:** | **ANNOTATION/COMMENT (STEP #):** |
|---|---|

**Solve system of equations for desired variable (4)**

$$f = \mu n$$

$$-(\mu n) + P\cos\theta = ma$$

Substituted for the friction force $f$, in the first term.

$$n = \underbrace{mg - P\sin\theta}$$

$$-\mu(mg - P\sin\theta) + P\cos\theta = ma$$

Eliminating the normal force and solving for the friction coefficient.

$$-\mu(mg - P\sin\theta) = ma - P\cos\theta$$

$$-\mu = \frac{ma - P\cos\theta}{mg - P\sin\theta}$$

$$\boxed{\mu = \frac{P\cos\theta - ma}{mg - P\sin\theta}}$$

Final expression for the friction coefficient

**Check units (5)**

$$[\mu] = \frac{\left[\dfrac{P\cos\theta - ma}{mg - P\sin\theta}\right]}{}$$

Note: Keeping the outer square brackets in the numerator and denominator until the last step is appropriate and technically required. This is because they contain terms that are summed; it is not evident until the end that their units are the same.

$$= \frac{[P\cos\theta - ma]}{[mg - P\sin\theta]}$$

$$= \frac{[[P][\cos\theta] + [m][a]]}{[[m][g] + [P][\sin\theta]]}$$

$$= \frac{[(N)(1) - (kg)(m/s^2)]}{[(kg)(m/s^2) + (N)(1)]}$$

$$[\mu] = N/N = 1 \quad \checkmark$$

Units for friction coefficient are unitless as expected.

**Check limiting cases (5)**

$$a \to 0:$$

$$\mu = \frac{P\cos\theta - ma}{mg - P\sin\theta}$$

$$= \frac{P\cos\theta}{mg - P\sin\theta} - (0)$$

For smaller accelerations, $a \to 0$, friction coefficient must be greater, as expected.

$$\mu = \frac{P\cos\theta}{mg - P\sin\theta} \quad \checkmark$$

| SOLUTION: | ANNOTATION/COMMENT (STEP #): |
|---|---|

**SOLUTION:**

$\theta \to 0$:

$$\mu = \frac{P\cos\theta - ma}{mg - P\sin\theta}$$

$$= \frac{P\cos(0) - ma}{mg - P\sin(0)} = \frac{P(1) - ma}{mg - P(0)}$$

$$\mu = \frac{P - ma}{mg} \checkmark$$

$$\mu = \frac{P\cos\theta - ma}{mg - P\sin\theta}$$

$$= \frac{10\,N\cos\theta - (5.0\,kg)(1.2\,m/s^2)}{(5.0\,kg)(9.8\,m/s^2) - 10\,N\sin\theta}$$

$$= \frac{10\,N\cos\theta - 6.0\,N}{49\,N - 10\,N\sin\theta}$$

$$\boxed{\mu = \frac{10\cos\theta - 6.0}{49 - 10\sin\theta}}$$

$\theta \to 15°$:

$$\mu = \frac{10\cos(15°) - 6.0}{49 - 10\sin(15°)}$$

$$= \frac{10(0.966) - 6.0}{49 - 10(0.259)}$$

$$= \boxed{0.075}$$

$\theta \to 75°$:

$$\mu = \frac{10\cos(75°) - 6.0}{49 - 10\sin(75°)}$$

$$= \frac{10(0.259) - 6.0}{49 - 10(0.966)}$$

$$= \boxed{-0.087}$$

**ANNOTATION/COMMENT (STEP #):**

Note: The unit analysis step is normally done by inspection and not written down. However, for more complicated expressions, it is best to write it out.

For the angle going to zero, the friction coefficient must get larger to compensate if the acceleration is to remain constant.

**Evaluate final expression numerically, leaving the value of angle for last (6)**

Note: Another reason to delay inserting numerical values into the expressions until the end is if there are multiple values to calculate.

Intermediate calculation of final values for acceleration.

Final value for $\theta = 15°$

Final value for $\theta = 75°$

This value is *unphysical, and therefore this situation cannot exist* because friction coefficients can only be positive.

A negative value for $\mu$ would be equivalent to the friction force helping to move the block instead of slowing it down.

# EXAMPLE 2.5

A climber with a mass of 55 kg is hanging from the underside of a horizontal overhang, connected to two ropes that join in a knot. The angles
the ropes make with the underside of the overhang are 15° and 40°.
What is the tension in each of the ropes?

| SOLUTION: | ANNOTATION/COMMENT (STEP #): |
|---|---|
| | With forces mentioned or inferred, we know Newton's Second Law is appropriate. |
| $m_1$ = 55 kg, $\alpha$ = 40°, $\beta$ = 15°, $g$ = 9.8 m/s², Tensions = ? | **Identify the raw data of the problem and the key request of the problem statement (1)** |

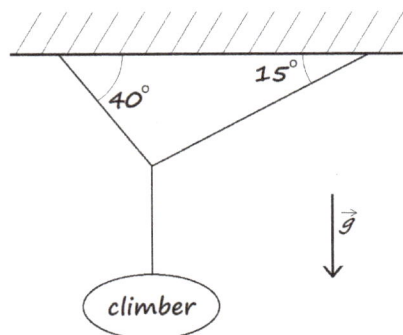

**Draw diagram to help identify all important parts (1)**

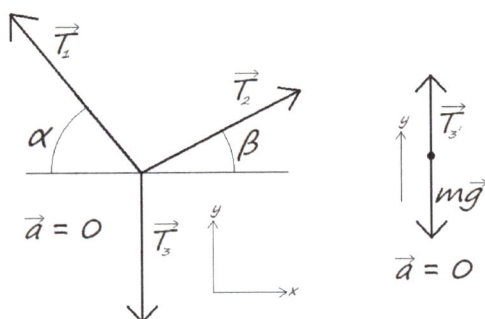

**Draw a free-body diagram and acceleration vector for each object and pick the appropriate coordinate system for each (1)**

Because they are on the same rope, the forces acting on the knot from below, and the climber from above are labeled 3 and 3′.

The acceleration is zero and is labeled as such instead of writing a vector.

Note: These coordinate systems do not have to be the same for each object.

## SOLUTION:

**For the climber:**

$$\sum \vec{F} = m\vec{a}$$
$$\vec{T}_{3'} + m\vec{g} = \vec{O}$$
$$(+T_{3'}\hat{j}) + (-mg\,\hat{j}) = (O\,\hat{j})$$

$$\boxed{T_{3'} - mg = O}$$

**For the knot:**

$$\sum \vec{F} = m\vec{a}$$
$$\vec{T}_1 + \vec{T}_2 + \vec{T}_3 = \vec{O}$$

$$(-T_1 \cos\alpha\,\hat{i} + T_1 \sin\alpha\,\hat{j})$$
$$+(+T_2 \cos\beta\,\hat{i} + T_2 \sin\beta\,\hat{j})$$
$$+(O\,\hat{i} - T_3\hat{j})$$
$$= (O\hat{i} + O\hat{j})$$

$$\hat{i}(-T_1 \cos\alpha + T_2 \cos\beta)$$
$$+\hat{j}(+T_1 \sin\alpha + T_2 \sin\beta - T_3)$$
$$= \hat{i}(O) + \hat{j}(O)$$

$$\boxed{\begin{array}{ll} x) & -T_1 \cos\alpha + T_2 \cos\beta = O \\ y) & +T_1 \sin\alpha + T_2 \sin\beta - T_3 = O \end{array}}$$

## ANNOTATION/COMMENT (STEP #):

**Translate vector equation $\sum \vec{F} = m\vec{a}$ for each object into scalar form (2)**

Note: The negative signs only start showing up in this step using unit vector representation.

The final equation for the climber

The knot has no mass, but this will be fine because the acceleration is zero.

For the knot, using a 2D coordinate system requires the use of sine and cosine functions to transform the vectors into unit vector form. Using trigonometric functions becomes easier with practice.

Note: The zero vector $\vec{O}$ is used intentionally as a reminder that both of the components on the right side are zero.

The terms are collected by component direction.

The final equations for the knot

## SOLUTION:

| variable | known? |
|----------|--------|
| $T_1$ | no |
| $T_2$ | no |
| $T_3$ | no |
| $T_{3'}$ | no |
| $\alpha$ | yes |
| $\beta$ | yes |
| $g$ | yes |
| $m$ | yes |

Unknowns = 4, Equations = 3

Same string:

$$\left|\vec{T}_3\right| = \left|\vec{T}_{3'}\right| \quad \text{or} \quad T_3 = T_{3'}$$

$$\boxed{\begin{aligned} T_3 - mg &= 0 \\ -T_1 \cos\alpha + T_2 \cos\beta &= 0 \\ +T_1 \sin\alpha + T_2 \sin\beta - T_3 &= 0 \end{aligned}}$$

Unknowns: $T_1$, $T_2$, $T_3$

$$T_3 = \underbrace{mg}$$

$$T_1 \sin\alpha + T_2 \sin\beta = \overbrace{(mg)}$$

$$T_1 \cos\alpha = T_2 \cos\beta$$

$$T_1 = T_2 \left(\frac{\cos\beta}{\cos\alpha}\right)$$

## ANNOTATION/COMMENT (STEP #):

**Check if the number of scalar equations equals the number of unknowns (3)**

Note: Very often, such a table and the listing of the number of unknowns and equations is not shown and is done in one's head instead. However, for more complicated problems, writing out such a table can be useful.

With 4 unknowns and only 3 equations, we will need at least 1 more equation or more data to solve these.

Extra equation that we need

Because the climber and knot are connected by the same rope, the force on each would have the same magnitude.

Final set of equations

Here, the substitution $T_3 = T_{3'}$ was executed, leaving 3 equations and 3 unknowns. The system of equations is now solvable.

**Use algebra to solve the system of equations simultaneously for parameter of interest (4)**

The result for $T_3$ is substituted into the third equation (in parentheses).

Solving the second equation for $T_1$.

## SOLUTION:

$$T_1 \sin \alpha + T_2 \sin \beta = mg$$

$$\left( T_2 \frac{\cos \beta}{\cos \alpha} \right) \sin \alpha + T_2 \sin \beta = mg$$

$$T_2 \left( \frac{\cos \beta}{\cos \alpha} \sin \alpha + \sin \beta \right) = mg$$

$$T_2 = \frac{mg}{\left( \dfrac{\cos \beta}{\cos \alpha} \sin \alpha + \sin \beta \right)}$$

$$T_1 = (T_2) \frac{\cos \beta}{\cos \alpha}$$

$$= \left( \frac{mg}{\left( \dfrac{\cos \beta}{\cos \alpha} \sin \alpha + \sin \beta \right)} \right) \frac{\cos \beta}{\cos \alpha}$$

$$= \left( \frac{\cos \beta}{\cos \alpha} \right) \frac{mg}{\left( \dfrac{\cos \beta}{\cos \alpha} \sin \alpha + \sin \beta \right)}$$

$$\boxed{\begin{array}{l} T_1 = \left( \dfrac{\cos \beta}{\cos \alpha} \right) \dfrac{mg}{\left( \dfrac{\cos \beta}{\cos \alpha} \sin \alpha + \sin \beta \right)} \\[2em] T_2 = \dfrac{mg}{\left( \dfrac{\cos \beta}{\cos \alpha} \sin \alpha + \sin \beta \right)} \\[1em] T_3 = mg \end{array}}$$

## ANNOTATION/COMMENT (STEP #):

Eliminating $T_1$ in the third equation to solve for $T_2$. The substitution is in the parentheses in the second line.

Note: The order of substitutions is chosen for convenience. No matter the order, all of the algebra would have worked out. If each algebra step is correct, then the same answer will always result.

Substituting the expression for $T_2$ into the expression for $T_1$ to get an expression in terms of the angles and *mg*.

Final expressions for each of the values of the tensions

## SOLUTION:

$$[T_1] = \left[\left[\frac{\cos\beta}{\cos\alpha}\right]\right] \cdot$$

$$\frac{[mg]}{\left[\left[\frac{\cos\beta}{\cos\alpha}\sin\alpha + \sin\beta\right]\right]}$$

$$= (1)\frac{(kg)\cdot(m/s^2)}{\left[\left[\frac{\cos\beta}{\cos\alpha}\sin\alpha\right] + [\sin\beta]\right]}$$

$$= \frac{kg\cdot(m/s^2)}{[(1+1)]} = \frac{kg\cdot(m/s^2)}{1}$$

$$= \left[\frac{kg\cdot m}{s^2}\right]\left[\frac{N}{kg\cdot m/s^2}\right]$$

$$[T_1] = N \checkmark$$

$$[T_2] = \frac{(kg)\cdot(m/s^2)}{\left[\left[\frac{\cos\beta}{\cos\alpha}\sin\alpha\right] + [\sin\beta]\right]}$$

$$= \frac{kg\cdot(m/s^2)}{[1+1]} = \frac{kg\cdot(m/s^2)}{1}$$

$$= \left[kg\frac{m}{s^2}\right]\left[\frac{N}{kg\cdot m/s^2}\right]$$

$$[T_2] = N \checkmark$$

$$[T_3] = [mg] = (kg)\left[\frac{m}{s^2}\right]$$

$$= \left[\frac{kg\cdot m}{s^2}\right]\left[\frac{N}{kg\cdot m/s^2}\right]$$

$$[T_3] = N \checkmark$$

## ANNOTATION/COMMENT (STEP #):

### Check units (5)

Checking if the units for each expression for the tensions is Newtons.

Note: The unit analysis step is usually not shown in submitted assignments. It is shown here to illustrate how it is done. Either way, unit analysis is important, as it is a good way to check your work for errors.

The units are as expected for $T_1$.

The units are as expected for $T_2$

The units are as expected for $T_3$

| SOLUTION: | ANNOTATION/COMMENT (STEP #): |
|---|---|

**Check physicalness (5)**

$$T_3 = mg = constant$$

Checking the tension for rope 3 is trivial as it is equal to a constant.

> Note: Checking predictions is often done in one's head or on scratch paper and is not shown on submitted work. However, it is important to check your work for errors.

Here, the checking will be in terms of the angles and not the lengths of the ropes which are unknown.

$$\underline{\beta \to 90°: \cos\beta \to 0 \text{ and } \sin\beta \to 1}$$

$$T_2 = \frac{mg}{\left(\dfrac{(0)}{\cos\alpha}\sin\alpha + 1\right)} = \frac{mg}{0+1} = mg$$

If $\beta \to 90°$, then rope 2 is vertical and is carrying the full weight of the climber; therefore, rope 1 does not contribute.

$$T_1 = \left(\frac{(0)}{\cos\alpha}\right)\frac{mg}{\left(\dfrac{(0)}{\cos\alpha}+1\right)} = \frac{0}{1} = 0$$

Recall, we do not know the lengths of the ropes; we only know the angles.

$$\underline{T_1 = 0 \checkmark}$$

The tension in $T_2$ becomes $mg$, and the tension $T_1$ becomes zero as expected

$$\underline{\alpha \to 90°: \cos\alpha \to 0, \sin\alpha \to 1}$$

$$T_2 = \frac{mg}{\left(\dfrac{\cos\beta}{(0)}(1) + \sin\beta\right)}$$

If $\alpha \to 90°$, then rope 1 is vertical and is carrying the full weight of the climber; therefore, rope 2 does not contribute.

$$= \frac{mg}{(\infty + \sin\beta)}$$

$$= \frac{mg}{\infty}$$

$$\underline{T_2 = 0 \checkmark}$$

The tension in rope 2 becomes zero as expected

## SOLUTION:

$$T_1 = \left(\frac{\cos\beta}{\cos\alpha}\right)\frac{mg}{\left(\frac{\cos\beta}{\cos\alpha}\sin\alpha + \sin\beta\right)}$$

$$= \frac{mg}{\left(\frac{\cos\alpha}{\cos\beta}\right)\left(\frac{\cos\beta}{\cos\alpha}\sin\alpha + \sin\beta\right)}$$

$$= \frac{mg}{\left(\sin\alpha + \sin\beta\left(\frac{\cos\alpha}{\cos\beta}\right)\right)}$$

$$= \frac{mg}{\left((1) + \sin\beta\left(\frac{(0)}{\cos\beta}\right)\right)}$$

$$= \frac{mg}{\left(1 + (0)\right)}$$

$$T_1 = mg \checkmark$$

$$T_3 = mg$$
$$= (55kg)\left(9.8\frac{m}{s^2}\right)$$
$$= 539 \left(\frac{kg \cdot m}{s^2}\right)\left(\frac{N}{kg \cdot m/s^2}\right)$$

$$\boxed{T_3 = 540\,N}$$

$$\left(\frac{\cos\beta}{\cos\alpha}\right) = \frac{\cos 15^\circ}{\cos 40^\circ} = \frac{(0.966)}{(0.766)} = 1.261$$

## ANNOTATION/COMMENT (STEP #):

Checking the tension in rope 1 for $\alpha \to 90^\circ$.

The expression for $T_1$ is rearranged so that there is only one term with $\cos\alpha$.

Note: It is often helpful to rearrange an expression to make it more convenient to evaluate the limiting behavior.

The tension in rope 1 becomes $mg$ as expected.

### Evaluate final expression numerically (6)

Note: The cancellation and collecting of the units as part of the numerical evaluation act as a final check of the units.

Final numerical answer for $T_3$

The value of $(\cos\beta/\cos\alpha)$ is computed here as a preliminary step to simplify subsequent calculations.

Note: For intermediate calculations, extra significant digits are usually retained.

| SOLUTION: | ANNOTATION/COMMENT (STEP #): |
|---|---|

$$T_2 = \frac{mg}{\left(\dfrac{\cos\beta}{\cos\alpha}\sin\alpha + \sin\beta\right)}$$

Computing value for $T_2$.

$$= \frac{540\,N}{(1.261)\sin 40° + \sin 15°}$$

$$= \frac{540\,N}{(1.261)(0.643) + (0.259)}$$

$$= 505\,N$$

$$\boxed{T_2 = 510\,N}$$

Final value for $T_2$

$$T_1 = \left(\frac{\cos\beta}{\cos\alpha}\right)\frac{mg}{\left(\dfrac{\cos\beta}{\cos\alpha}\sin\alpha + \sin\beta\right)}$$

Computing value for $T_3$.

$$= (1.261)(505\,N) = 637\,N$$

$$\boxed{T_1 = 640\,N}$$

Final value for $T_3$

# Solving Work-Kinetic Energy Problems

3

## Introduction

Although Newton's Second Law relates force, mass, and acceleration, it is limited because unless the force is constant, all we can conclude is what the acceleration would be at the instant those forces are being applied. Furthermore, in order to quantify the cumulative effect of a force acting on some object, we would have to invoke the relationships from basic kinematics to determine the distance traveled and what the change in speed might be. This is even harder if the force is not constant in magnitude or changes direction while it travels along a path.

What would be nice is to have some way to capture the effect of a force acting on an object as it moves along a path. Thanks to the tools of calculus, this can be done. We can do this by summing the effect of the force along each segment of the object's path. The result of applying calculus is that a relationship between the initial and final speeds of an object can be related directly to the total effect of a force acting on a body along a path.

This relationship, called the Work-Kinetic Energy Theorem, is a powerful one and is a steppingstone to the concept of the law of the conservation of energy. Not surprisingly, how an object responds to interactions cannot depend on how we describe it, whether it be with Newton's Second Law or the Work-Kinetic Energy Theorem. However, there will be occasions where one approach will be more natural or convenient than another. The derivation of the Work-Kinetic Energy Theorem is intimately tied to Newton's Second Law and can be argued to be an alternate formulation of it in terms of initial and final speeds, forces, and paths, instead of the instantaneous relationship between forces and acceleration of Newton's Second Law.

### Skills and Knowledge Needed to Solve Work-Kinetic Energy Problems

◊ *Conceptual understanding*:

- What an integral is and what it means to say that the value of an integral is equal to "the area under the curve"

- How to express the cumulative effect of a force that is non-constant or acting on an object from varying directions

- What a dot-product is

◊ *Familiarity with concepts, Kinetic Energy, Work, and Newton's Second Law*:

- Kinetic Energy: $K = \frac{1}{2}mv^2$
- Work: $W = \vec{F} \cdot \vec{r}$ or $W = \int_i^f \vec{F} \cdot d\vec{r}$
- Newton's Second Law: $\sum \vec{F} = m\vec{a}$

◊ *Word-Problem Skills*:

- How to re-express physical and mathematical concepts expressed in words as mathematical expressions

- Comfort with solving a system of equations

## TARGETS AND GOALS

In this chapter, you will learn the following:

✔ **How to identify if the Work-Kinetic Energy Theorem is appropriate to use.**

✔ **How to apply the Work-Kinetic Energy Theorem and use it to set up the mathematical expression that will describe the system's behavior.**

✔ **The usefulness and limitations of the Work-Kinetic Energy Theorem.**

## The Nature of Work-Kinetic Energy Problems

The Work-Kinetic Energy Theorem can be thought of as a before/after version of Newton's Second Law. Whereas Newton's Second Law relates the instantaneous acceleration of an object to the forces acting at that instant and the object's mass, the Work-Kinetic Energy Theorem relates how the speeds at some earlier and later time are related to the cumulative effect of the sum of the applied forces over the path taken by the object.

   Clues that the Work-Kinetic Energy Theorem could be used to describe what is happening to an object include if the information provided or desired is in the form of the initial and final speeds, the forces applied, and information regarding the path taken (distance or displacement).

**Work-Kinetic Energy Theorem**

$$\Delta K = K_f - K_i = \sum W \text{ where } \sum W = \int_i^f \vec{F}_{net} \cdot d\vec{r} \text{ and } K = \tfrac{1}{2}mv^2$$

*The change in kinetic energy of a body has the same value as the total work done on it, where the total work done is the cumulative effect of all of the forces acting along the path taken.*

## Caveats and Subtilties of the Work-Kinetic Energy Theorem

Understanding and using the mathematical expression for the Work-Kinetic Energy Theorem requires the use of calculus and geometry, which can seem complicated. However, this is less daunting if the mathematical relationship is seen as a short-hand version of the conceptual description.

   Specifically, the integral $\int_i^f \vec{F}_{net} \cdot d\vec{r}$ should be read as *"the cumulative effect of all of the forces acting along the path taken."*

   For most problems encountered in a first-year physics course, actual integration will not be necessary due to problems that either have constant forces or areas that can be integrated by inspection by breaking up the single area into sub-areas that have shapes such as triangles and rectangles that then can be easily computed and then summed.

   The other mathematical short-hand used in quantifying the net effect of the forces applied is the dot-product. The dot-product mathematically captures that for an object moving along a fixed path, say railroad tracks, the only component of the force that could cause a change in speed is the component of that force pointing along the allowed path.

   Below is a brief summary of these concepts that are covered in regular physics textbooks but included here within the context of their use in problem-solving. This summary may also be of use to students who find that the description in their textbook is initially not clear to them.

### The Effect of a Force Acting Along a Path; The Dot-Product

A way into understanding the Work-Kinetic Energy Theorem is to ask the following question:

   *What would be the expected change in speed due to the cumulative effect of a constant force acting on an object along some fixed path?*

   From Newton's Second Law, $\sum \vec{F} = m\vec{a}$, we would expect that only the component of the net force $\sum \vec{F} = \vec{F}_{net}$, that is along the fixed path could have any effect on the speed. For example, if the net force is directed along the path, then it can fully affect the speed of the object, but if it were perpendicular to the path along which the object is moving, then it would not be able to change the object's speed. This can be seen in the diagram below:

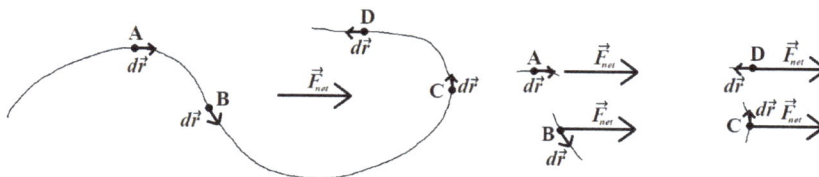

At each of the points along the path **A**, **B**, **C**, and **D**, the infinitesimal displacement vector, $d\vec{r}$, is drawn, showing its direction along the path at each point. If at each of the locations, the net force is from left to right as indicated by $\vec{F}_{net}$, then the effect of that force will be different depending on the relative directions of the possible displacement vector and that force.

The effect at each point is summarized below:

- **A**: the effect of the force would be to increase the speed along the path.

- **B**: the effect is to increase the speed only by a small amount as only a component of the force is parallel to the path.

- **C**: the force is perpendicular and cannot cause a change in the motion along the path.

- **D**: the force is directed opposite to the direction of motion along the path, so the effect would be to slow the object down.

Mathematically, the effect of this force times the displacement can be represented by $dW = F_{net,\parallel}\,dr$, where $dW$ is the work done by the force along the differential path segment of length $dr$ and $F_{net,\parallel}$ is the magnitude of the component of the net force along the path. Using trigonometry, this expression can be rewritten as $dW = (F_{net}\cos\theta)dr$ where $\theta$ is the angle between the vectors $\vec{F}_{net}$ and $d\vec{r}$. The mathematical short-hand for this is called the dot-product and is represented as $dW = \vec{F}_{net} \cdot d\vec{r}$.

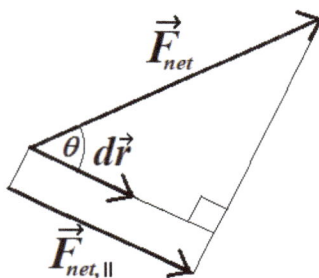

## Visual Evaluation of the Work Integral: "The Area Under the Curve"

The evaluation of the cumulative effect of a net force acting along a path is often represented as a plot of *the magnitude of that force that is parallel to the path at each point* as a function of the *distance along that path*. Visually, the cumulative effect of this force along the path would be the area enclosed by the curve of the force between the beginning and endpoints. This is shown in the figures below that will look familiar to anyone who has opened a physics textbook that discusses the Work-Kinetic Energy Theorem.

The first figure shows how the area enclosed by the curve of the force as a function of distance can be approximated by filling the area with many narrow rectangles, with the area enclosed by the curve being approximated by the sum of the areas of the rectangles.

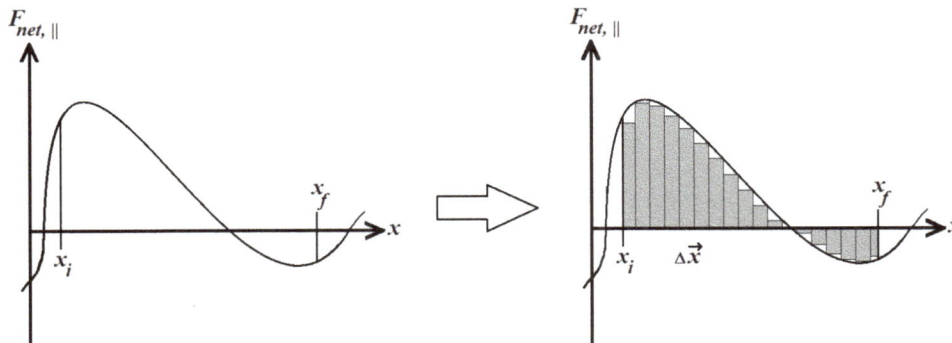

The important caveat here is that the common expression that an integral is equal to the "area under the curve" can be misleading. *Strictly speaking, the value of an integral can also be negative.* In the figure below, the dark-gray area has a positive area, and the light-grey area has a negative area. The net work, the sum of the areas, is positive because, here the positive area is larger than the area that has the negative value.

In terms of the relative directions of the applied net force and the path traveled, a positive net force on the graph of $F_{net,\parallel}$ vs. $x$ corresponds to it having a component pointing in the positive $x$-direction. For travel in the positive $x$-direction, such a force would cause the object to accelerate in the positive $x$-direction. In contrast, a net force that is negative is one with a component pointing in the negative $x$-direction and for travel in the positive $x$-direction, would cause the object to decelerate, i.e., accelerate in the negative $x$-direction.

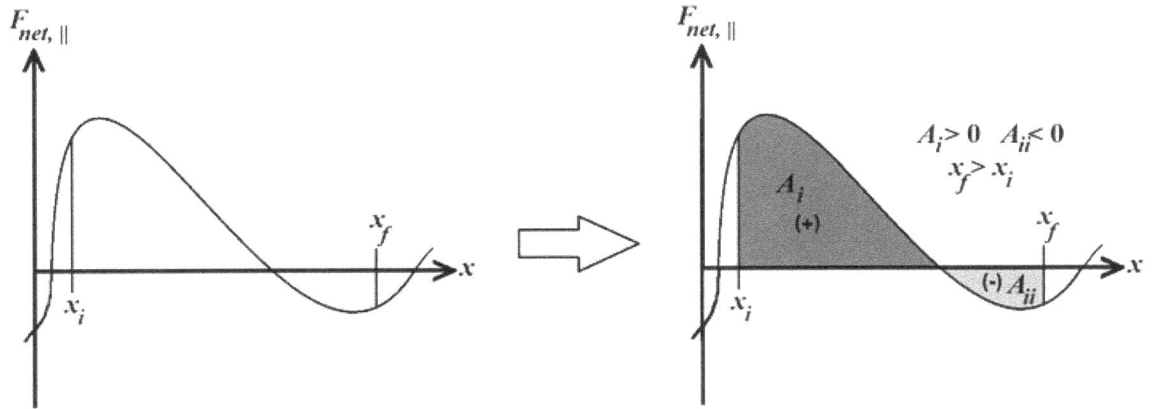

A common mistake students make is to take "the area under the curve" to mean the area between the highest point and lowest point. This is not correct, as illustrated below.

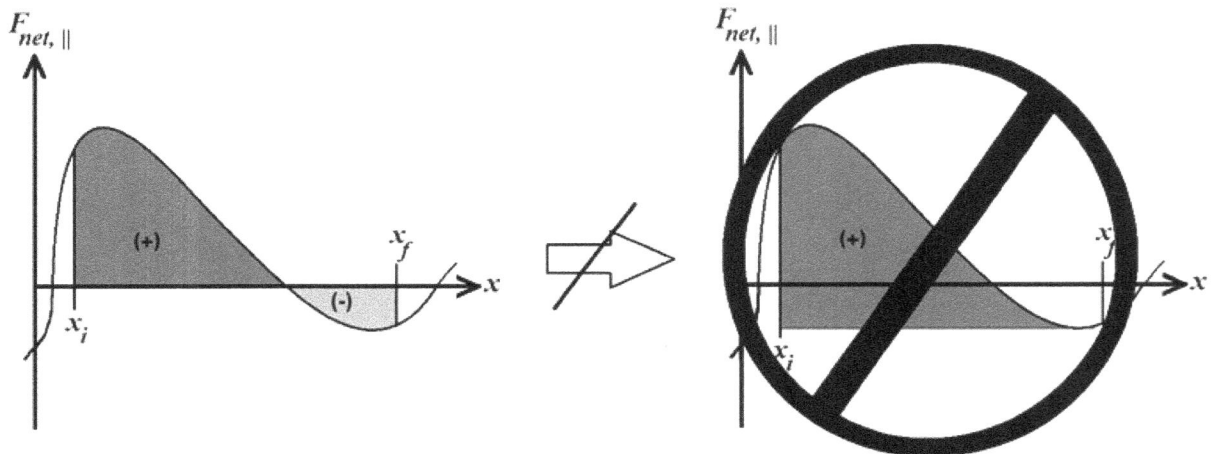

In practice, the actual evaluation of these integrals is often simplified by the boundaries of the areas in the force versus position plots being straight lines. This allows the integration to be done by subdividing the areas into triangles and rectangles, as illustrated below.

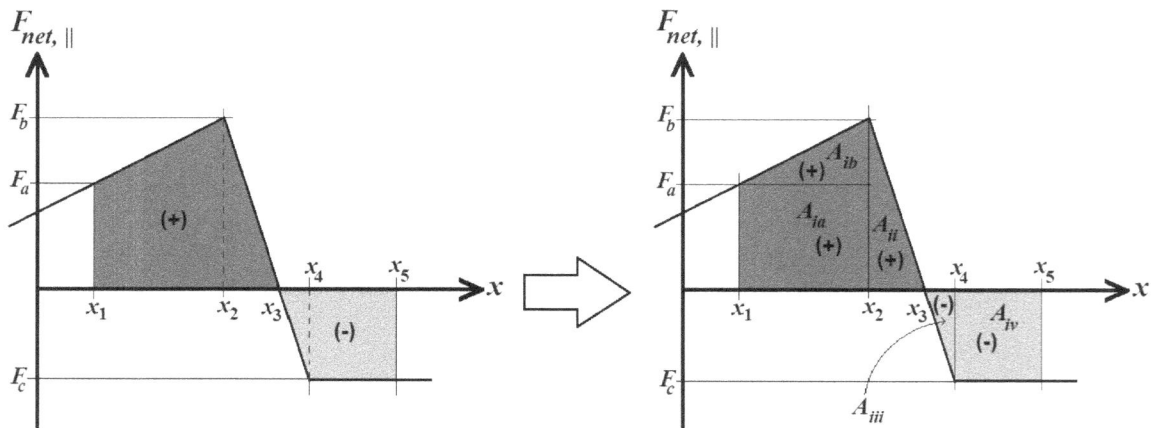

The figure above shows how these areas can be subdivided into smaller areas, with areas that are easier to compute and then sum. The areas in dark gray have positive areas, and those in light grey are negative.

$$\sum W = W_{1\to2} + W_{2\to3} + W_{3\to4} + W_{4\to5}$$

$$\sum W = [A_{ia} + A_{ib}] + A_{ii} + A_{iii} + A_{iv}$$

There is one more aspect to consider: the relative value of the final and initial positions. For example, if over the range of $x$ so that $x_f > x_i$ and $F(x) > 0$, then $W = \int_{x_i}^{x_f} F(x)dx$ will be positive. However, if the positions of the initial and final coordinates were reversed so that $x_f < x_i$ (i.e., the object traveled in the opposite direction), then the amount of work done would be negative. The following figures illustrate the importance of keeping the direction in mind when doing these problems.

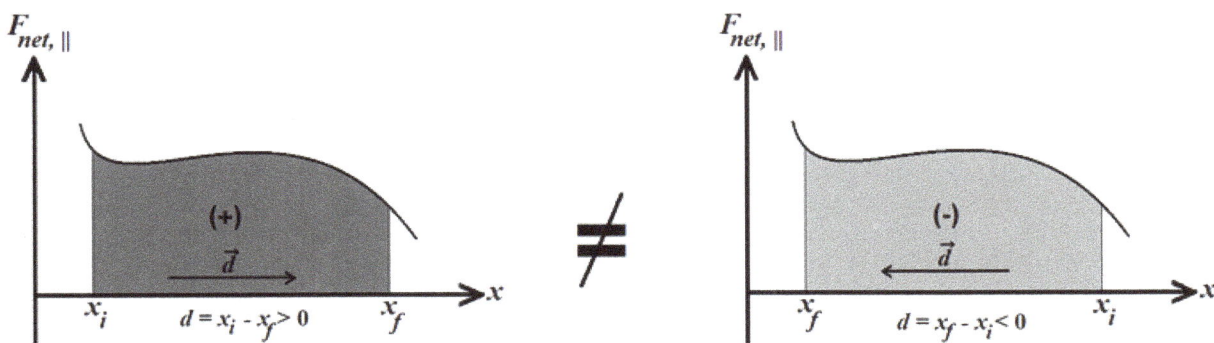

## The Problem-Solving Steps for Work-Kinetic Energy Problems

In general, problem-solving all begins with seeing the big picture: in other words, what category or "flavor" of physical situation is it? Is it an initial/final type of problem? Is there a conserved quantity involved? Is it an instantaneous problem, such as a force or torque? Instead of jumping to some previously derived formula, the power in starting with one of the key concepts of mechanics is that you can then use it as a starting point to create the relationship needed for the situation at hand. This will be good practice for later when the problems you will be solving, whether as a scientist or engineer, are ones for which there are no previously derived solutions.

### Identify the Type of Problem

As for all problem-solving, for homework, exams, or in your career, the first step is to identify the physical ingredients and physics ingredients. This will guide you to the principle that is best suited to describe the interactions and allow you to create mathematical relationships that will predict how the system will evolve.

### Hallmarks of a Work-Kinetic Energy Problem

- There are identifiable initial and final speeds for the object on which the forces are acting.
- There are forces described as acting over a distance along a path. Time is not explicitly noted.

### Recall the Physics Principle: Work-Kinetic Energy

In its mathematical form, the Work-Kinetic Energy relation is easy to write down. However, as always, one must correctly understand what it represents, hence the conceptual form below.

- *Mathematical Form:* $\sum W = \Delta K$ where $\sum W = \int_{\vec{r}_i}^{\vec{r}_f} \left( \sum \vec{F} \right) \cdot d\vec{r}$ , $K_f = \frac{1}{2}mv_f^2$, $K_i = \frac{1}{2}mv_i^2$ and $\Delta K = K_f - K_i$

- *Conceptual Form*: *The work done by the forces acting on a body moving along a path has the same value as the change in that body's kinetic energy as it traveled that path.*

To use a physics concept such as the Work-Kinetic Energy Theorem, one must know what it means and its origin. The same is true for all of the key physics principles available to describe interactions among objects. To this end, although we mostly use the mathematical form of the Work-Kinetic Energy Theorem in practice, a wise student will know what it means conceptually. This will ensure that you use it correctly and do not make any mistakes when you apply it.

**In general, it is not enough to be able to write down the equation form for any physics principle. To know how to use it, you must also be able to write down a conceptual, word version of the principle. Being able to do that is good practice as it will ensure you know how to use the concept correctly.**

## Use a Picture to Describe the Path of the Object and the Forces Acting on the Body

Here we draw a diagram for the initial and final states along with a free-body diagram and the displacement vector to aid in the initial setup of the problem. Alongside will be written the data for the parameters of the system.

## Write Down the General Principle to be Used and Then Adapt It to the Specific Situation

For Work-Kinetic Energy problems, there will only ever be one object that can have a change in its kinetic energy, but there can be several forces doing work. Thus, it is in the expression of the sum of the work, $\Sigma W$, where the specifics of the problem will mostly be found because every Work-Kinetic Energy problem will have an initial and a final kinetic energy term. For the case where there are two forces doing work on an object, the first two lines would look like this: $\Delta K = \Sigma W$ and $K_f - K_i = W_1 + W_2$. Note that the work terms are summed. It is when the expressions for each one are inserted, that the negative signs for those forces that do negative work will show up.

## Apply Any Other Needed Relationships to be Able to Solve for Parameter of Interest

After writing down the mathematical form for the Work-Kinetic Energy theorem, other equations that are appropriate are now written. Common examples would be to use a free-body diagram to identify all of the forces acting on the object, if the force of friction is doing any work, then $f = \mu n$ would be written, or, from the application of Newton's Second Law, the magnitude of the normal force that will be needed in the expression for the magnitude of the friction force.

Finally, in addition to the free-body diagram, draw the displacement vector defined within the same coordinate system as used for the free-body diagram. It is here that specific expressions for each of the work terms would be inserted. These explicit terms will include any negative signs that indicate whether the work due to a particular force (e.g., friction) is negative. Being explicit with the negative signs here will prevent errors from being made.

## With Any Additional Relationships Applied, Do the Algebra to Solve for the Parameter of Interest

Doing all the algebra symbolically rather than by inserting numerical values now will make combining all the mathematical relationships easier. It will also reveal if any values might cancel out and not be needed, which often occurs with friction problems.

## Check Units and Expectations

*The next steps are seldom shown in solutions for submitted assignments or exams, but they should still be done. They are often easy to do by inspection, or "in one's head," and will reveal if any errors were made. If any errors are found, it is easier to stop now and go back and find the error rather than going forward.*

Checking units is straightforward: if the final desired expression is expected to be for a speed, the units of the expression should be m/s.

Checking the predictions of the expression is harder but well worth doing. For instance, based on our intuition, an expression for a final speed can be expected to have a structure so that if the friction coefficient were to be set to zero, then the final speed would be greater. If, instead, the expression predicted that the final speed would be smaller, then there is likely an error in the previous steps, and it should be found before moving on.

## Put in Numerical Values to Find Answer and Check that Answer Is Physical

The final step is to put in the values provided for any of the parameters of the system and then check if the final value makes physical sense. Examples of "making sense" would if the final speed of an object was less than the speed of light or if a friction coefficient were negative. Neither of those is physically possible. A more subtle example is that because the expression for the kinetic energy is the product of terms that are all positive, then a kinetic energy term must always be positive. So, if an expression predicts that a kinetic energy must be negative, then either the solution has an error or the parameters being used correspond to an un-physical situation.

# Work-Kinetic Energy Problem-Solving Steps

<u>Hallmarks</u>: There are identifiable initial and final speeds for the object on which forces are acting, and the forces described are acting over a distance along a path or some displacement. Time is not explicitly noted.

**Work-Kinetic Energy Theorem**

$$\sum W = \Delta K = K_f - K_i$$

where $\sum W = \int_{\vec{r}_i}^{\vec{r}_f} \left( \sum \vec{F} \right) \cdot d\vec{r}$, $K_f = \frac{1}{2} m v_f^2$, and $K_i = \frac{1}{2} m v_i^2$

## Steps

1. **Identify components of system\*.** (A sketch is very useful for this.)

   - *Draw a diagram* for the initial and final states, along with a free-body diagram and the displacement vector to aid in the initial setup of the problem.

   - *Define the key variables* used with analytic expressions, including zeros for any coordinate systems.

2. **Adapt the general principle to the specific situation to be described\*.**
   - *For the Work term*, identify if the net force is constant, or, if it acts at a fixed angle to the path of the object.

      - If force is not constant or if the angle of force to the path is not constant, use the integral form for work

      - If force is constant and if the angle between force and path is constant, use $W = \vec{F} \cdot \vec{d}$

   - *For Kinetic Energy terms*, fill in the expression for kinetic energy, $\frac{1}{2} m v^2$, for initial and final states. Use zero for either that corresponds to the object being at rest.

3. **Identify and apply any other important relationships**.
   - *Use Newton's Second Law* as necessary to find relationships between magnitudes of forces (e.g., find an expression for the normal force so that the friction force, $f = \mu n$, can be expressed properly).

   - *Fill in any special values or relationships* for variables, for example, $y_f = h$, $v_i = 2v_f$, or simplifications of the dot-product.

4. **Do algebra and/or integration to solve equations for the parameter of interest.**
   - *Use only variables*, not numerical values.

5. **Check "physicalness" of the final symbolic relationship.**
   - *Perform unit analysis*.

   - *Check predictions* of limiting behavior.

6. **Evaluate expression using data** (if provided).
   - *Check physicalness and reasonableness* of the answer.

\* *These are the key steps for this type of problem.*

# EXAMPLE 3.1

A car accelerates from rest to a speed of 25 m/s over a distance of 500 m. The car's mass is 1,500 kg. What was the magnitude of the force applied by the engine? Assume the force was constant.

**SOLUTION:**

**ANNOTATION/COMMENT (STEP #):**

Change in speed, forces acting over a distance: Work-Kinetic Energy approach is appropriate to use.

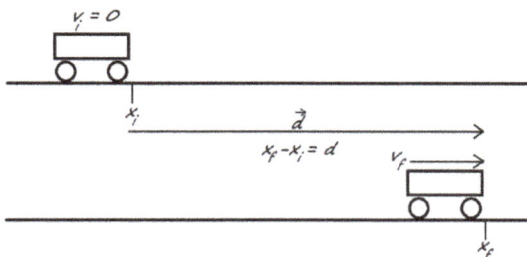

**Using diagram to identify key components of system (1)**

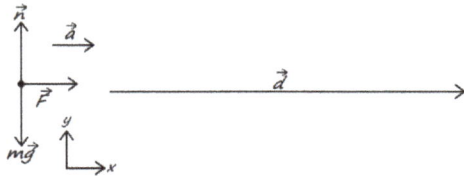

**Free-body-diagram for identifying directions of forces and acceleration (1)**

$v_i = 0 \text{ m/s}, v_f = 25 \text{ m/s},$

$m = 1,500 \text{ kg}, d = 50 \text{ m}$

$F = ?$

**Listing provided data and variable for which to solve (1)**

Here, only a single force is doing work to cause a change in speed.

$\sum W = \Delta K$

$W_{\vec{F}} = K_f - K_i$

**Applying key principle to problem (2)**

$\vec{F} \parallel \vec{d} \text{ so } \theta = 0 \text{ and}$

$W_{\vec{F}} = \vec{F} \cdot \vec{d} = +Fd$

Force and displacement vectors are constant, therefore can use

$W_{\vec{F}} = \vec{F} \cdot \vec{d} = Fd\cos(\theta)$

| SOLUTION: | ANNOTATION/COMMENT (STEP #): |
|---|---|

**Inserting supplemental expressions (3)**

$$W_{\vec{F}} = K_f - K_i$$
$$\vec{F} \cdot \vec{d} = \tfrac{1}{2}mv_f^2 - 0$$
$$+Fd = \tfrac{1}{2}mv_f^2$$

$v_i = 0$ and therefore $K_i = 0$.

**Doing algebra (4)**

$$F = \frac{mv_f^2}{2d}$$

Final expression for the force.

**Checking units (5)**

$$[F] = \left[\frac{mv_f^2}{d}\right] = \frac{[m]\big[v_f^2\big]}{[d]}$$

Note: This step is normally done in one's head or "by inspection," but even if not written down, it is important to do because it can help with finding errors early in the problem-solving process.

$$= \frac{(kg)\big(m/s\big)^2}{m} = kg\,\frac{m\!\!\!/}{m\!\!\!/ \cdot s^2}$$

$$= kg\,\frac{m}{s^2}$$

$$[F] = N \;\checkmark$$

Units are in Newtons as expected

$$\boxed{F = \frac{mv_f^2}{2d}}$$

**Checking predictions of final expression (5)**

To reach the same speed over a longer distance, a smaller force would be required.

$$F \propto \frac{1}{d} \;\checkmark$$

To go faster in the same distance, a larger force would be needed.

$$F \propto v_f^2 \;\checkmark$$

Note: Checking predictions of final expressions can normally be done by inspection or in one's head but should be done to catch errors early.

$$F = \frac{mv_f^2}{2d}$$

**Numerically evaluating final expression (6)**

$$= \frac{(1{,}500\ kg)(25\ m/s)^2}{2(500\ m)}$$

$$\boxed{F = 950\ N}$$

Value for force required.

950 N is equivalent to about 250 lbs of force, so this is not unreasonable.

# EXAMPLE 3.2

A car is accelerated by a force that increases linearly with the distance from zero to a maximum of 2,100 N at 100 m. The car then experiences a constant force in the opposite direction with a magnitude of 300 N for 20 m. The mass of the car is 1,700 kg. If it starts from rest, how fast is the car going at the end of the 120 m?

**SOLUTION:**

**ANNOTATION/COMMENT (STEP #):**

Forces, acting over a distance with changes in speed: Work-Kinetic Energy approach is appropriate to use.

$m = 1,700 \text{ kg}, v_i = 0, v_f = ?$

$F_{max} = 2,100 \text{ N}, F_{rev} = 300 \text{ N}$

**Identifying key components of system (1)**

$x_1 = 100 \text{ m}, x_2 = 120 \text{ m}$

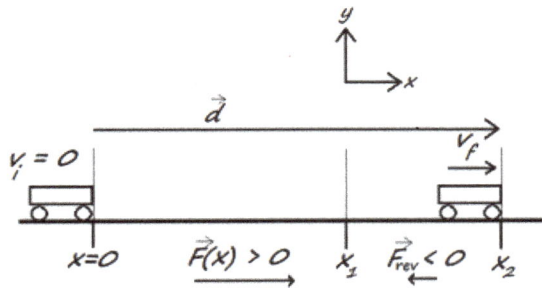

**Diagram of motion and forces showing their relative directions. (1)**

Here a variable force is doing work on the cart.

**Application of key principle (2)**

$$\sum W = \Delta K$$

$$\sum \left( \int_i^f \vec{F} \cdot d\vec{x} \right) = K_f - K_i$$

Because the force is not constant, the integral version of the expression for the work done by a force must be used.

$$\int_0^{x_1} F \, dx + \int_{x_1}^{x_2} F \, dx = \tfrac{1}{2} m v_f^2 - 0$$

The integral is split into two parts, one for each different force that is applied over the respective ranges for each.

## SOLUTION:

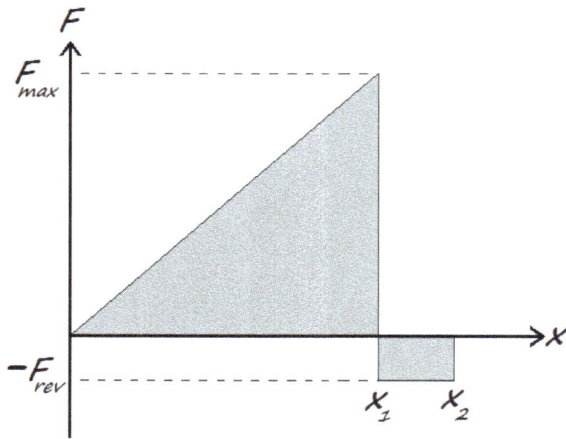

## ANNOTATION/COMMENT (STEP #):

### Plot of the force as a function of distance (2)

Recall, for parts of the plot for which the force is negative, the value of the "area" is negative.

$x = 0$ to $x = x_1$:

$$\int_0^{x_1} F\, dx = \tfrac{1}{2} F_{max}(x_1 - 0)$$

$$= \frac{F_{max}\, x_1}{2}$$

### Use supplemental information to simplify the expression (3)

The work for the sub-area of the plot of $F(x)$ from $x = 0$ to $x_1$. Because it is a triangle, the area can be expressed using $A = \tfrac{1}{2}\, bh = \tfrac{1}{2}\, F_{max} x_1$.

$x = x_1$ to $x = x_2$:

$$W_{x_1 - x_2} = \int_{x_1}^{x_2} F\, dx = (-F_{rev})(x_2 - x_1)$$

$$= -F_{rev}(x_2 - x_1)$$

The work for the sub-area of the plot of $F(x)$ from $x = x_1$ to $x_2$. Being a rectangle, the area can be expressed using. $A = bh = F_{rev}(x_2 - x_1)$.

Notice that the "height" here is negative, and the base is positive. Therefore, the area is "negative." Physically, negative work will tend to reduce speed.

$$\int_0^{x_1} F\, dx + \int_{x_1}^{x_2} F\, dx = \tfrac{1}{2} mv_f^2$$

$$\left(\frac{F_{max}\, x_1}{2}\right) + (-F_{rev}(x_2 - x_1)) = \tfrac{1}{2} mv_f^2$$

$$\frac{F_{max}\, x_1}{2} - F_{rev}(x_2 - x_1) = \tfrac{1}{2} mv_f^2$$

**SOLUTION:**

$$\tfrac{1}{2} m v_f^2 = \frac{F_{max} x_1}{2} - F_{rev}(x_2 - x_1)$$

$$v_f^2 = \left(\frac{2}{m}\right)\left(\frac{F_{max} x_1}{2} - F_{rev}(x_2 - x_1)\right)$$

$$v_f^2 = \frac{F_{max}}{m} x_1 - 2\frac{F_{rev}}{m}(x_2 - x_1)$$

$$v_f = \left(\frac{F_{max}}{m} x_1 - 2\frac{F_{rev}}{m}(x_2 - x_1)\right)^{1/2}$$

$$[v_f] = \left(\left[\frac{F_{max}}{m} x_1\right] + \left[\frac{F_{rev}}{m}(x_2 - x_1)\right]\right)^{1/2}$$

$$= \left(\frac{N}{kg} m + \frac{N}{kg} m\right)^{1/2} = \left(\frac{N}{kg} m\right)^{1/2}$$

$$= \left(\frac{(kg \cdot m/s^2)}{kg} m\right)^{1/2}$$

$$= \left(\frac{\cancel{kg}}{\cancel{kg}} \frac{m^2}{s^2}\right)^{1/2} = \frac{m}{s}$$

$$\boxed{v_f = \left(\frac{F_{max}}{m} x_1 - 2\frac{F_{rev}}{m}(x_2 - x_1)\right)^{1/2}}$$

$$v_f \sim \left(1/m\right)^{1/2}\left(F_{max} - F_{rev}\right)^{1/2} \checkmark$$

$$v_f \sim (x_1 - (x_2 - x_1))^{1/2} \checkmark$$

**ANNOTATION/COMMENT (STEP #):**

**Algebra to solve for final speed (4)**

Expression for final speed.

**Check units (5)**

Note: Checking units will seldom be included in full solutions, but it is important to do so to catch any errors early in the problem-solving process. Normally this can be done by inspection or in one's head.

Units are m/s as expected for speed.

**Check predictions for final expression (5)**

Note: This step can be done by inspection or in one's head. It should always still be done, even if not shown in solutions.

Would expect a greater final speed for:
- a smaller mass
- larger $F_{max}$, or smaller $F_{rev}$

Would expect a greater final speed for:
- larger $x_1$
- smaller $(x_2 - x_1)$

**SOLUTION:**

**ANNOTATION/COMMENT (STEP #):**

**Numerical evaluation of the final expression. (6)**

$$v_f = \left| \frac{F_{max}}{m} x_1 - 2\frac{F_{rev}}{m}(x_2 - x_1) \right|^{1/2}$$

$$= \left| \frac{2,100 \text{ N}}{1,700 \text{ kg}} 100 \text{ m} \right.$$

$$\left. -2\frac{300 \text{ N}}{1,700 \text{ kg}}(120 \text{ m} - 100 \text{ m}) \right|^{1/2}$$

$$= \left| 124\frac{m^2}{s^2} - 7.06\frac{m^2}{s^2} \right|$$

$$\boxed{v_f = 11\,\frac{m}{s}}$$

Final speed is about 25 mph, so not an unreasonable answer.

# EXAMPLE 3.3

A cart is constrained to roll on a straight track. The cart experiences a constant force that varies from directly behind to from the side and then from ahead. Find an expression for the final speed as a function of the magnitude of the force $F$, the mass of the cart $m$, the angle of the force relative to the track $\theta$, the distance traveled $d$, and the initial speed of the cart $v_i$.

Evaluate this expression for $F = 0.10$ N, $m = 1.2$ kg, $v_i = 0.50$ m/s, $d = 1.0$ m, and for the force coming from 30° from directly behind, 70° from directly behind, and from 45° from ahead.

**SOLUTION:**

$v_i = 0.50\,\text{m/s},\ \left|\vec{F}\right| = F = 0.10\ \text{N},$

$d = 1.0\,\text{m},\ m = 1.2\ \text{kg}$

$\theta = 30°, 70°, 135°$

$v_f = ?$

*Top View*

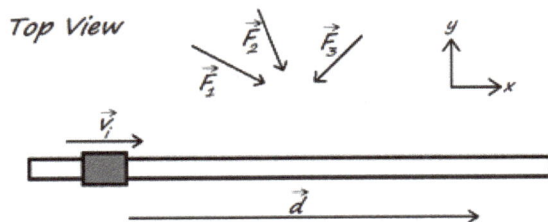

**ANNOTATION/COMMENT (STEP #):**

Force acting on an object that is changing speed: Work-Kinetic Energy approach is appropriate to use.

**Listing data to be used and what is to be found (1)**

**Draw diagram to help identify the geometry of the problem (1)**

## SOLUTION:

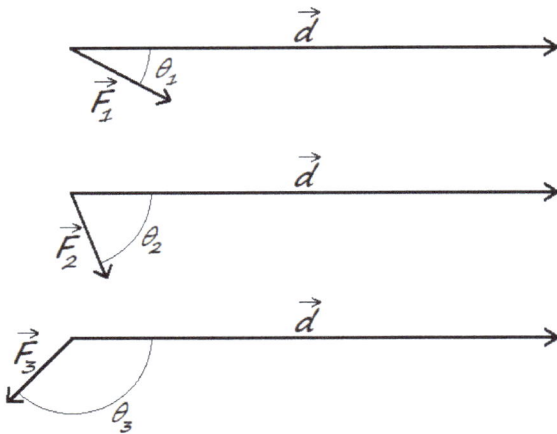

## ANNOTATION/COMMENT (STEP #):

**Second diagram to assist in the evaluation of the work done by the force (1)**

Expectations:

- Force 1 would tend to accelerate the cart.
- Force 2 will tend to accelerate the cart too but to a lesser extent than force 1.
- Force 3 would tend to decelerate the cart.

$$\Delta K = \sum W$$
$$K_f - K_i = W_{\vec{F}}$$

**Application of key principle (2)**

$$W_{\vec{F}} = \vec{F} \cdot \vec{d} = Fd\cos\theta$$

Because the force is constant, use $W = \vec{F} \cdot \vec{d} = Fd\cos(\theta)$.

$$\frac{1}{2}mv_f^2 - \frac{1}{2}mv_i^2 = \vec{F} \cdot \vec{d}$$
$$\frac{1}{2}mv_f^2 - \frac{1}{2}mv_i^2 = Fd\cos\theta$$

**No other supplemental expressions to put into expression (3)**

$$\frac{1}{2}m(v_f^2 - v_i^2) = Fd\cos\theta$$

**Using algebra to solve for final speed (4)**

$$v_f^2 - v_i^2 = \frac{2Fd\cos\theta}{m}$$

$$v_f^2 = \frac{2Fd\cos\theta}{m} + v_i^2$$

$$v_f = \left( \frac{2Fd\cos\theta}{m} + v_i^2 \right)^{1/2}$$

Expression for the final speed

## SOLUTION:

$$[v_f] = \left[\frac{2Fd\cos\theta}{m} + v_i^2\right]^{1/2}$$

$$= \left[\left[\frac{2Fd\cos\theta}{m}\right] + [v_i^2]\right]^{1/2}$$

$$= \left[N\frac{m}{kg} + \left(\frac{m}{s}\right)^2\right]^{1/2}$$

$$= \left[\left[\frac{kg\cdot m}{s^2}\right]\frac{m}{kg} + \left(\frac{m}{s}\right)^2\right]^{1/2}$$

$$= \left[\frac{m^2}{s^2} + \left(\frac{m}{s}\right)^2\right]^{1/2} = \left[\left(\frac{m}{s}\right)^2\right]^{1/2}$$

$$= \frac{m}{s}$$

$$\boxed{v_f = \left(\frac{2Fd\cos\theta}{m} + v_i^2\right)^{1/2}}$$

$$v_f \sim (\cos\theta + constant)^{1/2} \checkmark$$

$$v_f \sim (F + constant)^{1/2} \checkmark$$

For $v_f \in \mathbb{R}$, then
$$\left(\frac{2Fd\cos\theta}{m} + v_i^2\right) \overset{!}{>} 0$$

and $v_i^2 \overset{!}{>} \dfrac{2Fd\cos\theta}{m}$ $\checkmark$

## ANNOTATION/COMMENT (STEP #):

**Checking units (5)**

Note: Although this step should always be done, it seldom needs to be included in any solutions. However, it is an easy way to check if an error has been made.

Units are m/s as expected.

**Check predictions of final relationship (5)**

Note: This step should always be done but can normally be done in one's head to check for errors.

The greater the force along the direction of the track, the greater the final speed.

For greater force, there is greater change in final speed.

For the value of the final speed to be real and physical, the expression within the parentheses must be positive, and thus the force cannot be too large. Otherwise, the cart would go backward, which is inconsistent with the problem setup.

Limits found on the applicability of this final expression

## SOLUTION:

$$v_f = \left[ \frac{2Fd\cos\theta}{m} + v_i^2 \right]^{1/2}$$

$$= \left[ \frac{2(0.10\,N)(1.0\,m)\cos\theta}{1.2\,kg} \right.$$

$$\left. + \left( 0.50\,\frac{m}{s} \right)^2 \right]^{1/2}$$

$$= (0.167\cos\theta + 0.250)^{1/2}\,\frac{m}{s}$$

| $\theta$ | $\cos\theta$ | $v_f$ |
|---|---|---|
| 30° | +0.866 | 0.63 m/s |
| 70° | +0.342 | 0.55 m/s |
| 135° | −0.707 | 0.36 m/s |

## ANNOTATION/COMMENT (STEP #):

**Evaluation of expression numerically (6)**

Creation of simplified equation so that only one variable is left to insert.

Table for final speeds as a function of angle.

Final answers for $v_f$ are as expected:

$v_{f1} > v_{f2} > v_i$ and $v_{f3} < v_i$

# EXAMPLE 3.4

A roller coaster cart, m = 800 kg, is at the top of a hill traveling forward at 5.0 m/s. It then rolls down the hill traveling a distance along the track of 72 m while dropping in height 50 m. At the bottom, it is traveling at a speed of 29 m/s. What was the total work done by the forces of friction and air drag as it traveled down the hill? If the net force from friction and drag were constant, what would be its value?

## SOLUTION:

## ANNOTATION/COMMENT (STEP #):

Forces applied along a path with changes in speed: Work-Kinetic Energy approach is appropriate to use.

$m = 800\,kg,\ v_i = 5.0\,m/s,$

$v_f = 29\,m/s$

$h = 50\,m,\ d = 72\,m$

let $y_f = 0$ and $y_i = h$

$W_R = ?,\ \left|\vec{R}\right| = R = ?$

**Listing data and what is to be found (1)**

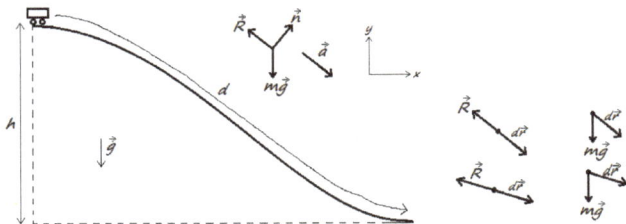

**Identifying key components of system with a diagram. Free-body diagram added as a reminder of the two forces acting. (1)**

$$\Delta K = \sum W$$
$$K_f - K_i = W_g + W_R$$
$$\tfrac{1}{2}mv_f^2 - \tfrac{1}{2}mv_i^2 = W_g + W_R$$
$$\frac{m}{2}\left(v_f^2 - v_i^2\right) = W_g + W_R$$

**Application of key principle (2)**

Work done by gravity and resistive forces (air-drag and friction)
→ Thus 2 work terms

| SOLUTION: | ANNOTATION/COMMENT (STEP #): |
|---|---|

**Identifying proper expression for the work terms (2)**

$$W_g = -mg\,\Delta y$$

Because gravity is a conservative force, only the end points of the change in position matter.

$$\frac{m}{2}(v_f^2 - v_i^2) = (-mg\,\Delta y) + W_R$$

$$= -mg(y_f - y_i) + W_R$$

**Identify supplemental information to use (3)**

$$y_i = h,\ y_f = 0$$

Definitions for initial and final heights.

**Doing algebra (4)**

$$\frac{m}{2}(v_f^2 - v_i^2) = -mg(0 - h) + W_R$$

$$= mgh + W_R$$

$$W_R = \frac{m}{2}(v_f^2 - v_i^2) - mgh$$

<u>Final expression solved for work done by the friction and drag forces.</u>

**Checking units (5)**

Note: This step is not normally done but should always be done at least in one's head.

$$[W_R] = \left[\frac{m}{2}(v_f^2 - v_i^2) - mgh\right]$$

$$= \left[\frac{m}{2}(v_f^2 - v_i^2)\right] + [-mgh]$$

$$= kg\left(\left(\frac{m}{s}\right)^2 - \left(\frac{m}{s}\right)^2\right) + kg\,\frac{m}{s^2}m$$

$$= kg\,\frac{m^2}{s^2} + kg\,\frac{m^2}{s^2}$$

$$= kg\,\frac{m^2}{s^2}$$

$$[W_R] = J \checkmark$$

<u>Units are Joules as expected.</u>

**SOLUTION:**

$$\boxed{W_R = \frac{m}{2}(v_f^2 - v_i^2) - mgh}$$

$R = \text{resistive force}, \therefore W_R \overset{!}{<} 0$

$W_R \sim C - mgh,$

$W_R \text{ decreases with } h \checkmark$

$W_R = \frac{m}{2}(v_f^2 - v_i^2) - mgh$

$$= \frac{800 \, kg}{2}\left(\left(29\frac{m}{s}\right)^2 - \left(5.0\frac{m}{s}\right)^2\right)$$

$$- (800 \, kg)\left(9.8\frac{m}{s^2}\right)(50 \, m)$$

$= -65.6 \, kJ$

$W_R = -66 \, kJ$

$R = ? \text{ (assume constant)}$

$W_R = -Rd$

$R = -\dfrac{W_R}{d}$

$$= -\frac{(-66 \, kJ)}{(72 \, m)}$$

$\boxed{R = 920 \, N}$

**ANNOTATION/COMMENT (STEP #):**

**Check predictions of final expression (5)**

Note: This step can be done by inspection. Doing so is a good way to catch sign errors.

Expect the work from friction and drag to be negative.

For fixed initial and final speeds, the work done by the resistive forces would have to be more negative to compensate if released from a higher point.

**Numerically evaluate final expression (6)**

Final value for work done by resistive forces.

**Value is negative as expected. (6)**

Next solving for value of resistive force.

**Using $W_R = -Rd$ because $\vec{R}$ is always pointing in the opposite direction of the instantaneous displacement. (3)**

**Doing algebra (4)**

Average magnitude of the resistive force.

920 N is about 207 lbs of force. Not an unreasonable value.

# EXAMPLE 3.5

A railroad car hits a large spring, slowing it down from 4.5 m/s to 2.0 m/s in a distance of 50 cm. What must have been the spring rate for the spring? The mass of the railroad car is 25×10³ kg.

## SOLUTION:

$v_i = 4.5 \, m/s, \; v_f = 2.0 \, m/s$

$m = 25 \times 10^3 \, kg, \; \Delta x = 25 \, cm$

Define: $x_i = 0$ and $x_f = 0.25 \, m$

$k = ?$

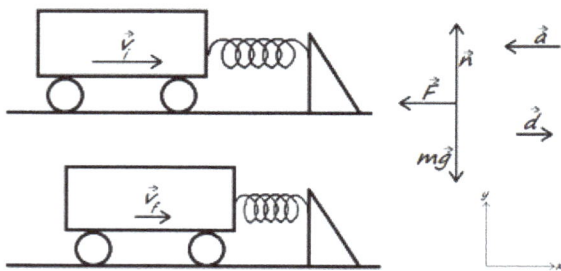

$$\Delta K = \sum W$$
$$K_f - K_i = W_{spr}$$
$$\tfrac{1}{2} m v_f^2 - \tfrac{1}{2} m v_i^2 = \tfrac{1}{2} k \left( x_i^2 - x_f^2 \right)$$

## ANNOTATION/COMMENT (STEP #):

Forces applied along a path with changes in speed: Work-Kinetic Energy approach is appropriate to use.

**Writing the data and what is being desired to be known (1)**

**Figure to help identify all components of the system (1)**

**Applying key principle to the problem (2)**

Only one force is doing work to change the speed of the railroad car.

Using expression for work done by a spring $W_{sp} = \tfrac{1}{2} k (x_i^2 - x_f^2)$.

| SOLUTION: | ANNOTATION/COMMENT (STEP #): |
|---|---|

**Incorporate supplemental information (3)**

$$\frac{1}{2}mv_f^2 - \frac{1}{2}mv_i^2 = \frac{1}{2}k\left(x_i^2 - x_f^2\right)$$

$$\frac{1}{2}mv_f^2 - \frac{1}{2}mv_i^2 = \frac{1}{2}k\left(0 - x_f^2\right)$$

$x_i = 0$.

**Doing algebra to solve for spring rate (4)**

$$\frac{\cancel{1}}{\cancel{2}}m\left(v_f^2 - v_i^2\right) = -\frac{\cancel{1}}{\cancel{2}}kx_f^2$$

$$m\left(v_f^2 - v_i^2\right) = -kx_f^2$$

$$k = m\frac{\left(v_i^2 - v_f^2\right)}{x_f^2}$$

Note the swapping of the order of the speed terms when the negative sign is canceled.

**Check units (5)**

$$[k] = \left[m\frac{\left(v_i^2 - v_f^2\right)}{x_f^2}\right]$$

$$= [m]\frac{\left([v_i^2] + [v_f^2]\right)}{\left([x_f^2]\right)}$$

$$= kg\frac{(m^2/s^2)}{(m^2)} = kg\frac{m^2}{s^2\,m^2}$$

$$= \left(kg\frac{m}{s^2}\right)\frac{1}{m} = \frac{N}{m}$$

Note: This step is seldom shown explicitly but should always be done as an easy way to check the work and catch any errors before moving on.

Units are N/m as expected.

**Checking predictions of final expression (5)**

$$\boxed{k = m\frac{\left(v_i^2 - v_f^2\right)}{x_f^2}}$$

$k \sim 1/x^2$ ✓

$k \sim \left(v_i^2 - v_f^2\right)$ ✓

$k \sim m$ ✓

Note: This step should always be done to find any errors and can be done by inspection, even if not written out.

To stop at a shorter distance would require a stiffer spring.

For a greater change in speed, a stiffer spring would be required.

The greater the mass of the railroad car, the greater the spring rate would need to be.

The expression properly predicts a positive value for the spring constant.

**SOLUTION:**

$$k = m \frac{(v_i^2 - v_f^2)}{x_f^2}$$

$$= (25 \times 10^3 \, kg)$$

$$\cdot \frac{\left(4.5 \, \tfrac{m}{s}\right)^2 - \left(2.0 \, \tfrac{m}{s}\right)^2}{(0.50 \, m)^2}$$

$$= (25 \times 10^3) \left(\frac{16.3}{0.25}\right) \left(kg \, \frac{m^2}{s^2}\right) \left(\frac{1}{m^2}\right)$$

$$= 1.6 \times 10^6 \, \frac{N}{m}$$

$$= 1.6 \times 10^6 \, \frac{N}{m} \left(\frac{1 \, m}{1,000 \, mm}\right)$$

$$\boxed{k = 1.6 \, \frac{kN}{mm}}$$

**ANNOTATION/COMMENT (STEP #):**

Numerical evaluation of final expression

Final value.

The value of the spring rate is comparable to those used on railroad car suspensions.

# Solving Conservation of Energy Problems

**4**

## Introduction

The concept of energy is one of the most powerful in physics. It is one of the few conserved quantities in nature, yet the amount of energy in a system is something we can only infer. Unlike forces and accelerations that we can sense with our bodies through our sense of touch (forces) and with our inner ear (accelerations), the only form of energy that can be felt is thermal (radiation).

Thus, using energy and the principle of the Conservation of Energy requires us to define quantities for which we have no direct sense. To illustrate this, one only needs to ask the following questions: Can you feel the amount of energy associated with the speed of the car in which you are riding? Can you sense the amount of energy associated with being at the top of a slide? For both, the answer is "no."

To use the principle of the Conservation of Energy, we must pay a small price by creating and using concepts that are a bit abstract. However, this price is worth paying because it provides a pathway to solving problems that Newton's Laws cannot and a way to solve problems that can be very easy. *We pay a small price for the abstraction required, but we gain much more by being able to use a new powerful tool to model interactions.*

### Skills and Knowledge Needed to Solve Conservation of Energy Problems

◊ *Word-Problem skills*:

- How to re-express physical and mathematical concepts expressed in words as mathematical expressions

- Comfort with solving a system of equations and doing algebra symbolically

◊ *Conceptual understanding*:

- How to identify interactions by conservative forces that will correspond to changes in potential energy

- How to identify interactions by non-conservative forces

◊ *Familiarity with concepts of Kinetic Energy, Work, and Potential Energy*:

- Kinetic Energy: $K = \frac{1}{2}mv^2$

- Work: $W = \vec{F} \cdot \vec{r}$ or $W = \int_i^f \vec{F}(\vec{r}) \cdot d\vec{r}$

- Potential Energy:

  - Gravity: $U = mgy$

  - Springs: $U = \frac{1}{2}kx^2$

## TARGETS AND GOALS

**In this chapter, you will learn the following:**

✔ **How to determine if the use of Conservation of Energy is appropriate.**

✔ **How to apply Conservation of Energy and use it to set up the mathematical expression that will describe the behavior of the system.**

✔ **The limits of applicability of the principle of Conservation of Energy.**

## The Nature of Energy and Conservation of Energy Law Problems

Conservation of Energy problems, and conservation law problems in general, can be considered initial/final or before/after problems in that the total amount of the thing being conserved (e.g., energy, momentum, or angular momentum) is unchanged or "conserved." One of the other characteristics of conservation law problems is that time is not explicitly part of the expressions used. As such, although "before/after" is often used, it is not strictly appropriate, as those words imply that there is a way to explicitly account for the amount of time that has passed. Therefore, in this book, "initial/final" will be used to describe the beginning-state and end-state of the system being investigated.

### To Use the Law of Conservation of Energy, It Is Important to Understand the Following

*For an isolated system, the amount of energy within that system will never change.*
   This seemingly simple sentence is loaded with powerful subtleties:

- What is meant by isolated? To be isolated means that the objects in the system do not interact with anything outside of the system. Recall that Newton's Third Law says for every interaction, there must be two things interacting. With that in mind, it can be said that for a system to be isolated, there is no object that is interacting with anything outside the system.[1]

- What types of energy are being conserved within the system? For the purpose of mechanics, there are three types: *Kinetic*, *Potential*, and *Non-conservative Work*.

   - Kinetic energy is the energy associated with the motion of a body. For each object in the system that is moving, it will have its own kinetic energy term: $K = \frac{1}{2}mv^2$.

   - Potential energy is the energy associated with the rearrangement of the bodies of the system (e.g., changing height, compression of a spring, and so on) associated with interactions that are conservative. For each interaction by conservative forces, we have one potential energy term (e.g., $U = mgy$ or $U = \frac{1}{2}kx^2$).

   - Non-conservative work is the energy associated with the work done by non-conservative forces. The most familiar example would be the work done by friction. Another example would be the work done by an engine or motor. In each case, there is an irreversible physical process involved, for example, the creation of heat by friction or the transformation of chemical energy into motion.

The first two types of energy, kinetic and potential, are forms of mechanical energy. This type of energy is associated with the motion and arrangement of bodies with respect to each other that interact via conservative forces. When only conservative forces are present, the transformation of kinetic energy to and from potential energy occurs with no losses.

But for the third type of energy, the work done by non-conservative forces, the interactions between bodies by such forces result in mechanical energy being transformed into heat[2] or heat being transformed into mechanical energy[3]. No matter the direction of transformation, it is a one-way transformation. As such, non-conservative work will either increase or decrease the amount of mechanical energy in a system.

For a course on mechanics, the question becomes, "What form would a mathematical expression for the conservation of energy take?" For situations where only conservative forces are acting, the Law of Conservation of Energy can be written $E_{mech,f} = E_{mech,f}$, where $E_{mech,f}$ is the sum of the final kinetic energies and potential energies, and $E_{mech,i}$ is the sum of the initial kinetic and potential energies where $E_{mech,f} = (\Sigma K_f + \Sigma U_f)$, $E_{mech,i} = (\Sigma K_i + \Sigma U_i)$.

---

1   Technically, when an object such as a stone is released and falls, the Earth is also accelerating upward as predicted by Newton's Third Law due to their mutual interaction by gravity; thus, initial and final kinetic energy terms for the Earth should be included. However, practically, the Earth's speed change can be neglected due to its much larger inertia and thus corresponding negligible acceleration as is predicted by Newton's Third Law. Because of this, the kinetic energy of the Earth does not change in a measurable way, and the kinetic energy terms for the Earth can be left out of our expressions. This would not be the case if the Earth were interacting with an object with a mass even only 1% of the mass of the Earth.

2   The most familiar example of heat creation is friction.

3   Heat can be transformed into kinetic energy in an engine.

To account for effects of non-conservative forces, we must adapt the fundamental equation by adding the contribution of non-conservative work on the *initial* side. This additional term represents additions to or subtractions from the initial mechanical energy by the non-conservative forces.

The following statement will be the starting point for all problems for which the Conservation of Energy is appropriate to use:

**Law of Conservation of Energy for Mechanics**

$$E_f = E_i + \Sigma W_{nc}$$

*The amount of mechanical energy in the system in its final state is equal to the amount of mechanical energy in its initial state plus any contributions (additions or subtractions) due to the work done by non-conservative forces.*

# Caveats and Subtilties of the Law of Conservation of Energy

## Path Independence/Potential Energy

Unlike mechanical properties that have a vector (directional) character, such as acceleration, velocity, or momentum, energy expressions have no explicit direction information contained in them. Instead, any such information is described in the problem set-up.

An important example that illustrates the concept of path independence would be that the answer to these two questions is the same:

- What is the speed of an object dropped from rest through a height $h$?

- What is the speed of an object attached to a massless string that is released from rest and swings down through a vertical distance $h$?

In the first case, the motion is down in a straight line, but in the second, the motion is along an arc. *When dealing with forces with which potential energies can be identified (i.e., conservative forces), the path taken does not matter.* Mathematically, the solution to both problems would be the same. Notice the use of "speed" instead of "velocity" to distinguish between the scalar value and the vector quantity.

The consequences for practical problem-solving are that the initial step of "abstracting" and the final step of "de-abstracting" the problem carry even more significance than for problems solved using Newton's Second Law for which the direction information is more closely tied to the final answers and is more explicit in the problem setup.

## Scalar Values Only

Closely related to the principle of path-independence is that *all representations of energy, kinetic, potential, or otherwise are scalar (i.e., an amount of each).* It is within the setup of the problem that the amounts of energy and their relationship to the positions of the objects or their paths of travel will be defined. As such, it is important to do this defining well, with no ambiguity.

## Forces of Constraint Do No Work

What is a force of constraint? This is a force that constrains something to follow a certain path. An example of this would be the normal force acting on a block due to the surface on which it slides. The surface constrains its path to be on the surface, but because the normal force is perpendicular to the path, the work done is zero; in other words, $W_n = \vec{n} \cdot \vec{d} = nd\cos 90° = nd(0) = 0$.

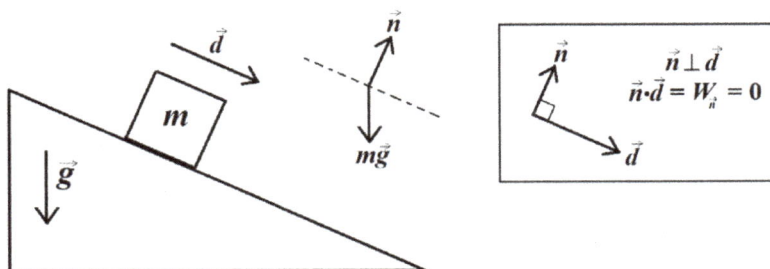

A second example is the tension in the string of a pendulum that constrains a mass to travel in an arc. For a pendulum, the force of tension is perpendicular (normal) to the path for each point along the path $d\vec{s}$. In other words, $dW_{\vec{T}} = \vec{T} \cdot d\vec{s} = (T)(ds)\cos 90^\circ = (T)(ds)(0) = 0$.

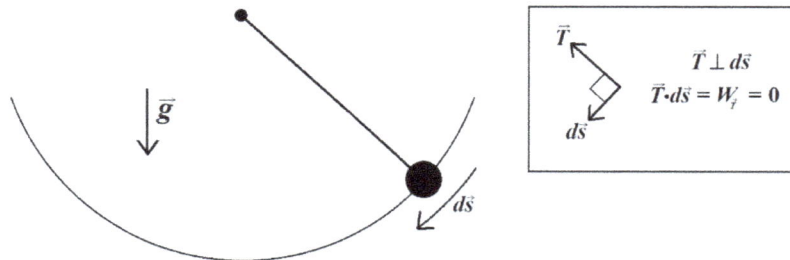

A third example of a force of constraint that does no net work would be a string that connects two blocks so that they move in tandem. This string constrains the motion so that they are coupled. For both blocks, the magnitude of the tension acting on each is the same while each undergoes the same displacement. However, although the displacement (the distance and direction) for each block will be the same, the tension force experienced by each will be in opposite directions. This results in the work done by the tension of the string on one block being exactly offset by the work done by the tension on the other block; in other words, for $\left|\vec{T}_{on\,1}\right| = \left|\vec{T}_{on\,2}\right| = T$ and $\left|\vec{d}_1\right| = \left|\vec{d}_2\right| = d$, we get the following net work:

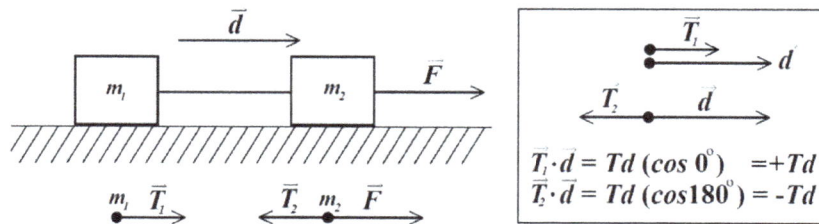

$$
\begin{aligned}
W_{T,\,net} &= W_{T\,on\,1} + W_{T\,on\,2} \\
&= (\vec{T}_1 \cdot \vec{d}) + (\vec{T}_2 \cdot \vec{d}) \\
&= Td(+1) + Td(-1) \\
&= Td - Td \\
&= 0
\end{aligned}
$$

## Non-Physical Mathematics

With the expression for the kinetic energy being a scalar expression ($K = \frac{1}{2}mv^2$), there is the possibility that the value for the kinetic energy could be less than zero, depending on the values of the other parameters of the system. How is this to be interpreted? Because all of the factors in the expression, $\frac{1}{2}$, $m$, and $v^2$ can only be greater than zero, any mathematical solution that would predict that the kinetic energy is less than zero is, therefore, a non-physical solution. *This is a reminder never to forget to check the "physicalness" of your answers and that physics is more than math and computation.*

## Initial/Final States

Although very often the language "before" and "after" is commonly used in conservation problems, it is important to understand that when something happens cannot always be inferred directly from the final expressions derived using the principle of conservation of energy. What is more accurate is to use "initial" and "final" to describe the state of the system in terms of the initial and final positions of each of the objects and their speeds.

*It is important to keep in mind that the states "initial" and "final" more exactly describe the change in the configuration of the system rather than the states "earlier in time" and "later in time," although the amount of time that has transpired can often be inferred.*

# The Problem-Solving Steps for Conservation of Energy Problems

As mentioned elsewhere in this book, "*A well-understood problem is a problem half-solved.*" When applied to problem-solving using the Conservation of Energy, this takes the form of identifying the system and the interactions within it. The number of interactions will determine the number of terms terms for the potential energy and the work done by the non-conservative forces, with each object that can move within the system having its own kinetic energy term.

*The setup for a Conservation of Energy problem is one of identifying proper energy terms to use to modify the fundamental equation $E_f = E_i + \Sigma W_{nc}$ for the particular problem to be solved.*

## Identify the Type of Problem

The first step with any problem is to identify a principle that can be used to solve the problem to create a final relationship that will describe how the system responds to the interactions and system parameters.

## Hallmarks of a Conservation of Energy Problem

- There are identifiable *initial* and *final* configurations or states for the system. This includes initial and final speeds, initial and final positions.

- The problem may include a description of a non-conservative force acting over a distance. An example of this would be the force of friction acting along the path of an object.

## Recall the Physics Principle: Conservation of Energy

- *Mathematical Form*: $E_f = E_i + \Sigma W_{nc}$ or $(\Sigma K_f + \Sigma U_f) = (\Sigma K_i + \Sigma U_i) + \Sigma W_{nc}$

- *Conceptual Form*: *The amount of mechanical energy of the system in its Final State is equal to the amount of mechanical energy in its Initial State plus any contributions (additions or subtractions) due to the Work done by Non-Conservative forces where the mechanical energy for a given state is the sum of all of the kinetic energy and potential energy of the system.*

To blindly use a principle without understanding it can result in non-physical predictions and little chance of discovering the source of the error. Whereas the mathematical relationship for the Conservation of Energy will be our starting point for solving problems, never forget what it represents.

**As with every physics principle to be used, understanding it conceptually is essential to being able to use it successfully, quantitatively, and mathematically.**

## Apply the Law of Conservation of Energy

Once the Law of Conservation of Energy has been identified as the concept that will be used, the type and number of each of the terms must be identified.

- For each object that could move, there will be an initial and a final kinetic energy term.

- For each conservative force of interaction, there will be one initial and one final potential energy term.

- For each non-conservative force doing work, there will be one term in the sum of the total work done by the non-conservative forces.

It is the setup of the problem where the physics is applied. This is the most important part of the problem. If it is setup correctly, then your final starting expression will be an accurate mathematical model for the system. It might not be very tidy but that is O.K.

## Simplification by Including Special Values and Relationships

Before any algebra is performed is a good time to include any special values in the expression. Examples include the following:

- If it is known that any of the initial or final speeds are zero, replacing the corresponding kinetic energy terms with zero can be done now.

- Since it is the change in position and not the absolute position that matters with potential energy, we are free to set convenient points in the system as our zeros for the coordinate systems. An example of this would be to set either the initial or final height, $y_i$ or $y_f$, to zero. With all possible zeros included, the final expression will be easier to solve.

## Doing the Algebra

Once the problem has been setup and simplified, then the initial physics is done, and the algebra begins. This is where the relationship is rearranged so that the parameter (variable) of interest is alone on one side of the equal sign.

## Re-Conceptualizing the Physics

Once the relationship has been rearranged as desired, check that the units are correct and that it conceptually makes the predictions we expect.

- Checking units is a good intermediate step that can reveal if the work done up to this point was done correctly. If the units are not as expected, say $m^2/s$ for speed, then there is no point in wasting time going forward with the rest of the problem.

- If the units are what is expected, then the final arrangement of the relationship can be evaluated to see if it makes predictions that are expected.

## Insertion of Final Parameters

Once the expression has been checked for its ability to make physical predictions, then the values for which it is to be evaluated can be inserted and the final value of the parameter of interest can be calculated. If you have done everything correctly, then the answer will be physical (e.g., speed is less than the speed of light) and the correct value.

# Conservation of Energy Problem-Solving Steps

Hallmarks: Work (forces and distances), initial and final speeds, initial and final positions are mentioned.

> **Law of Conservation of Energy for Mechanics**
>
> $$E_f = E_i + \Sigma W_{nc} \quad \text{or} \quad (\Sigma K_f + \Sigma U_f) = (\Sigma K_i + \Sigma U_i) + \Sigma W_{nc}$$

## Steps
1. **Identify components of system*** (A sketch is very useful for this)
    - *Count number of objects moving* (for changes in kinetic energy)
    - *Count number of objects changing position* (for changes in potential energy)
    - *Count number of non-conservative interactions* of objects with other objects (for work done on or by each object)
    - *Define the key variables* used with analytic expressions, including zeros for any coordinate systems
2. **Fill in placeholders for each energy term***
    - For each moving object, *add one initial and one final Kinetic Energy* term on *each* side of the equation using $K$ as appropriate
    - For each object changing position, *add one initial and one final Potential Energy term* on *each* side of the equation using $U$ as appropriate
    - For each non-conservative work term, *add one Work term* on *initial* side of equation
3. **For each energy term substitute the appropriate expression**
    - *Kinetic energy*:
        - If initial or final speed is zero, substitute zero
        - If initial or final speed is not zero, substitute $\frac{1}{2}mv^2$
    - *Potential energy*:
        - If an initial or final potential energy is zero, substitute zero
        - If an initial or final position is such that the potential energy is non-zero, substitute in full expression; e.g., $mgy$ or $\frac{1}{2}kx^2$
    - *Non-conservative work*:
        - For constant forces, $W = \vec{F} \cdot \vec{d} = Fd\cos\theta_{\vec{F},\vec{d}}$
        - For non-constant forces, $W = \int \vec{F} \cdot d\vec{r}$
        - If needed, use Newton's Second Law to find expression for the force $\vec{F}$
    - *Fill in any special values* for variables (e.g., $y_f = h$, $v_i = v_2$)
4. **Do algebra to solve equations for parameter of interest**
    - *Use only variables*, not numerical values
5. **Check "physicalness" of final relationship**
    - *Perform unit analysis*
    - *Check predictions* of limiting behavior
6. **Evaluate expression using data** (if provided)
    - *Check physicalness and reasonableness* of answer

*These are the key steps for this type of problem.*

# EXAMPLE 4.1

A ball is tossed upward at a speed of 1.1 m/s. What is its elevation relative to the point it is thrown when its final speed is 3.2 m/s?

**SOLUTION:**

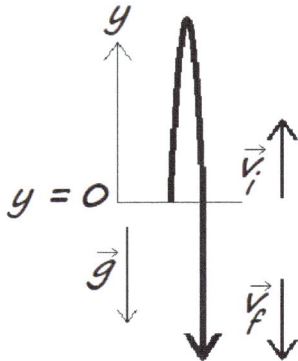

$$y = 0$$

$$\left(\begin{array}{c} 1 \text{ object} \\ \text{moving} \end{array}\right) \Leftrightarrow \left(\begin{array}{c} 1 \text{ K.E. term} \\ \text{per side of} \\ \text{equation} \end{array}\right)$$

$$\left(\begin{array}{c} 1 \text{ interaction by} \\ a \text{ conservative} \\ \text{force (gravity)} \end{array}\right) \Leftrightarrow \left(\begin{array}{c} 1 \text{ P.E. term} \\ \text{per side of} \\ \text{equation} \end{array}\right)$$

$$\left(\begin{array}{c} \text{No non-} \\ \text{conservative} \\ \text{work done} \end{array}\right) \Leftrightarrow (\text{no } W_{nc} \text{ terms})$$

**ANNOTATION/COMMENT (STEP #):**

Before/After with change in speed and arrangement of system: Energy approach is appropriate to use.

**Draw a picture to help identify all of the components of the system and the coordinate system to be used (1)**

Note: Although we are dealing with velocities, the Conservation of Energy approach only deals with speeds. The final answer will need to be checked for physicalness.

**Identify the number of each type of term to be used in general expression (1)**

Note: This step is normally not written and is instead done in one's head, although for very complicated problems it may be wise to write it out.

| SOLUTION: | ANNOTATION/COMMENT (STEP #): |
|---|---|

$$v_i \neq 0, v_f \neq 0$$
$$y_i = 0, y_f = ?$$

**Define variables being used for later simplification (1)**

$$E_f = E_i + \sum W_{nc}$$
$$K_f + U_f = K_i + U_i + 0$$
$$K_f + U_f = K_i + 0$$
$$K_f + U_f = K_i$$

**Start with fundamental relationship and fill in the appropriate energy terms (2)**

Note: When convenient, putting in values that are zero is appropriate at this step.

$$\tfrac{1}{2}mv_f^2 + mgy_f = \tfrac{1}{2}mv_i^2$$

**Next, put in specific expressions for each energy term (3)**

$$\tfrac{1}{2}mv_f^2 + mgy_f = \tfrac{1}{2}mv_i^2$$
$$\tfrac{1}{2}\cancel{m}v_f^2 + \cancel{m}gy_f = \tfrac{1}{2}\cancel{m}v_i^2$$
$$\tfrac{1}{2}v_f^2 + gy_f = \tfrac{1}{2}v_i^2$$
$$gy_f = \tfrac{1}{2}v_i^2 - \tfrac{1}{2}v_f^2$$

**Physics done, now do algebra (4)**

$$\boxed{y_f = \tfrac{1}{2g}(v_i^2 - v_f^2)}$$

Final expression for the variable of interest

This equation should look familiar as one of the 1D Kinematics formulas, as is expected for a single body moving up and down.

**Unit analysis (5)**

Note: Checking units is normally not shown in assignments that are turned in. It is shown here to illustrate how it is done. It is important to do unit analysis whether it is to be shown or not because it is a good way to check your work and catch your errors.

$$[y_f] = \left[\tfrac{1}{2g}(v_i^2 - v_f^2)\right]$$
$$= \left(1 \Big/ \left(\tfrac{m}{s^2}\right)\right)\left(\tfrac{m}{s}\right)^2$$
$$= \frac{s^2 \cdot m^2}{m \cdot s^2}$$
$$= \frac{\cancel{s^2} \cdot m^{\cancel{2}}}{m \cdot \cancel{s^2}}$$
$$[y_f] = m \checkmark$$

Units are as expected. Therefore, it is worth moving to the next step.

## SOLUTION:

$$y_f = \frac{1}{2g}(v_i^2 - v_f^2)$$

If $v_f^2 > v_i^2$, $y_f < 0$ ✓

and also

If $v_f^2 > v_i^2$, $y_f < 0$ ✓

$$y_f = \frac{1}{2g}(v_i^2 - v_f^2)$$

$$= \frac{1}{2\left(9.8\frac{m}{s^2}\right)}\left[\left(1.1\frac{m}{s}\right)^2 - \left(3.2\frac{m}{s}\right)^2\right]$$

$$= \frac{s^2}{19.6\ m}(1.21 - 10.24)\frac{m^2}{s^2}$$

$$= \frac{s^2}{19.6\ m}(-9.03)\frac{m^2}{s^2}$$

$$\boxed{y_f = -0.46\ m}$$

## ANNOTATION/COMMENT (STEP #):

### Analyze the expression for physicalness (5)

For $|v_f|$ larger than $|v_i|$, the final height will be below the initial height. This makes sense, as for heights above the initial height, the speed will first decrease to zero then increase again but never exceed the initial speed.

It is only for heights below the initial height that the speed can be expected to be larger than the initial speed.

Note: Checking predictions of the final expression is normally done in one's head but still should be done. Doing so can catch errors early so that they can be fixed before moving on.

### Evaluate final expression numerically (6)

Final answer.

This answer makes sense. For the final speed to be greater than the initial speed, the object will have to have fallen past the point at which it was launched upward.

# EXAMPLE 4.2

A simple (but large) pendulum consists of a mass of 2.2 kg on the end of a massless string of length 12.4 m. The pendulum is released from rest at an angle of 60° from the vertical. How fast is the mass traveling when it reaches the bottom of its swing?

**SOLUTION:**

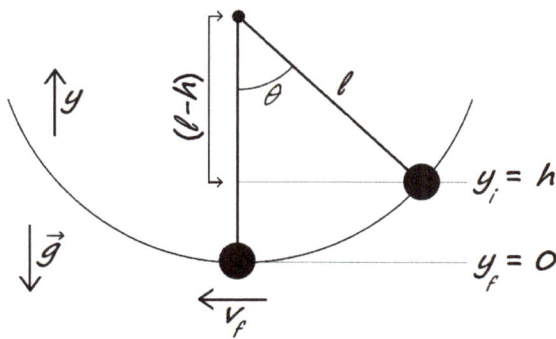

**ANNOTATION/COMMENT (STEP #):**

Before/After with change in speed and arrangement of system: Energy approach can be used.

Draw a picture to help identify all of the components of the system and the coordinate system to be used (1)

$$\begin{pmatrix} 1 \ object \\ moving \end{pmatrix} \Leftrightarrow \begin{pmatrix} 1 \ K.E. \ term \\ per \ side \ of \\ equation \end{pmatrix}$$

$$\begin{pmatrix} 1 \ interaction \ by \\ a \ conservative \\ force \ (gravity) \end{pmatrix} \Leftrightarrow \begin{pmatrix} 1 \ P.E. \ term \\ per \ side \ of \\ equation \end{pmatrix}$$

$$\begin{pmatrix} No \ non- \\ conservative \\ work \ done \end{pmatrix} \Leftrightarrow (no \ W_{nc} \ terms)$$

Identify the number of each type of term to be used in general expression (1)

Note: This step is normally done in one's head.

| SOLUTION: | ANNOTATION/COMMENT (STEP #): |
|---|---|

$$\text{let} \quad y_i = y_f + h$$
$$y_f = 0, \, y_i = h$$

$$v_i = 0, \, y_f = 0, \, y_i = h, \, v_f = ?$$

**Define variables being used for later simplification (1)**

$$E_f = E_i + \sum W_{nc}$$
$$K_f + U_f = K_i + U_i + 0$$
$$K_f + U_f = 0 + U_i$$
$$K_f + U_f = U_i$$
$$\tfrac{1}{2}mv_f^2 + mgy_f = mgy_i$$

**Start with fundamental relationship and fill in the appropriate energy terms (2)**

**Put in specific expressions for each energy term (3)**

$$\tfrac{1}{2}mv_f^2 + mg(0) = mg(h)$$
$$\tfrac{1}{2}mv_f^2 + 0 = mgh$$

**Put in variables based on the coordinate system being used (3)**

$$\tfrac{1}{2}mv_f^2 = mgh$$
$$\tfrac{1}{2}\cancel{m}v_f^2 = \cancel{m}gh$$
$$\tfrac{1}{2}v_f^2 = gh$$
$$v_f^2 = 2gh$$
$$v_f = \sqrt{2gh}$$

**Physics done, now do algebra (4)**

$$(\ell - h) = \ell\cos\theta$$
$$h = \ell - \ell\cos\theta$$
$$h = \ell(1 - \cos\theta)$$

Inserting expression for height in terms of angle based on geometry.

$$\boxed{v_f = \sqrt{2g\ell(1 - \cos\theta)}}$$

Final expression for the variable of interest.

Note: This answer is independent of mass. It is not uncommon for some variables to be canceled in the algebra step. This is one of the reasons why solving problems symbolically is recommended.

**SOLUTION:**

$$[v_f] = \left[\sqrt{2gh}\right]$$

$$= \sqrt{1\left(\frac{m}{s^2}\right)(m)}$$

$$= \sqrt{\left(\frac{m}{s}\right)^2}$$

$$[v_f] = \frac{m}{s} \checkmark$$

$$v_f = \sqrt{2gh} \propto \sqrt{h} \checkmark$$

$$v_f = \sqrt{2g\ell\,(1 - \cos\theta)}$$

$$= \sqrt{2\left(9.8\frac{m}{s^2}\right)(12.4\ m)(1-\cos 60°)}$$

$$= (122)^{1/2}\left(\left(\frac{m}{s}\right)^2\right)^{1/2}$$

$$\boxed{v_f = 11\frac{m}{s}}$$

**ANNOTATION/COMMENT (STEP #):**

**Unit analysis (5)**

Note: Unit analysis is normally done in one's head.

Units are m/s as expected for speed.

**Check that final expression is physical (5)**

For longer drop, final speed is larger as expected.

Note: Checking physicalness is normally done in one's head.

**Insert numerical values and evaluate expression (6)**

Final answer

Note: This is an example where the forces are conservative, and the answer is independent of path taken. As such, this would be the same speed if the object was simply dropped from rest.

# EXAMPLE 4.3

A mass of 0.25 kg is compressed against a spring by 0.20 m on a frictionless horizontal surface and released from rest. How fast is it moving when it loses contact with the spring? How fast would a block of mass 0.50 kg be moving if the spring were compressed by the same amount? The spring-rate of the spring is 2.0 N/m.

| SOLUTION: | ANNOTATION/COMMENT (STEP #): |
|---|---|
| | Before/after with change in speed and arrangement of system: Energy approach can be used. |

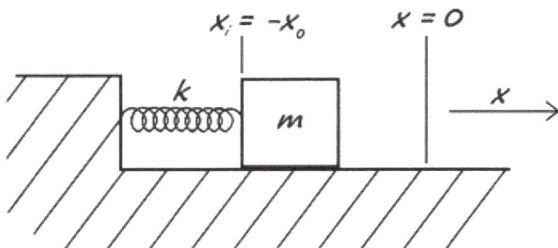

**Use diagram to define key variables and parameters (1)**

$$\begin{pmatrix} 1 \text{ object} \\ \text{moving} \end{pmatrix} \Leftrightarrow \begin{pmatrix} 1 \text{ K.E. term} \\ \text{per side of} \\ \text{equation} \end{pmatrix}$$

$$\begin{pmatrix} 1 \text{ interaction by} \\ \text{a conservative} \\ \text{force (spring)} \end{pmatrix} \Leftrightarrow \begin{pmatrix} 1 \text{ P.E. term} \\ \text{per side of} \\ \text{equation} \end{pmatrix}$$

$$\begin{pmatrix} \text{No non-} \\ \text{conservative} \\ \text{work done} \end{pmatrix} \Leftrightarrow (\text{no } W_{nc} \text{ terms})$$

**Identify the number of each type of term to be used in general expression (1)**

Note: This step is normally done in one's head.

| SOLUTION: | ANNOTATION/COMMENT (STEP #): |
|---|---|
| $m = 0.25$ kg, $0.50$ kg | **Define variables being used for later simplification (1)** |
| $x_i = -x_o = -0.20$ m, $x_f = 0.00$ m | |
| $k = 2.0$ N/m | Defining the original compression as $x_i = -x_0 = -0.20$ m. The negative sign will be important later. |
| $v_i = 0.00$ m/s, $v_f = ?$ | |

$$E_f = E_i + \sum W_{nc}$$
$$K_f + U_f = K_i + U_i + 0$$

**Start with fundamental relationship and fill in the appropriate energy terms (2)**

$$K_f + 0 = 0 + U_i$$
$$K_f = U_i$$

**Put in variables based on the coordinate system being used along with any terms equal to zero (3)**

$$\tfrac{1}{2}mv_f^2 = \tfrac{1}{2}kx_i^2$$

**Put in specific expressions for each energy term (3)**

**Do algebra (4)**

$$\tfrac{\cancel{1}}{\cancel{2}}mv_f^2 = \tfrac{\cancel{1}}{\cancel{2}}kx_i^2$$
$$v_f^2 = \frac{k}{m}x_i^2$$

Notice that taking a square root requires two solutions, + and –, be included.

$$\boxed{v_f = \pm\left(\sqrt{\frac{k}{m}}\right)(x_i)}$$

Final relationship between key parameters.

**Unit analysis (5)**

> Note: This step is normally done in one's head but is well-worth doing to confirm that the algebra is likely correct.

$$[v_f] = \left[\sqrt{\frac{k}{m}}x_i\right]$$
$$= \left(\sqrt{\frac{N/m}{kg}}\right)(m)$$
$$= \sqrt{\left(\frac{N}{kg \cdot m}\right)\left(\frac{kg\ m/s^2}{N}\right)}(m)$$
$$= \sqrt{\left(\frac{\cancel{N}}{\cancel{kg} \cdot \cancel{m}}\right)\left(\frac{\cancel{kg}\ \cancel{m}/s^2}{\cancel{N}}\right)}(m)$$
$$[v_f] = \frac{m}{s}\ \checkmark$$

Units are (m/s) as expected for speed.

**SOLUTION:**

**ANNOTATION/COMMENT (STEP #):**

**Check predictions of final relationship to expectations and physicalness (5)**

$$v_f = \left(\sqrt{\frac{k}{m}}\right)(|x_i|) = \left(\sqrt{\frac{k}{m}}\right)(x_o)$$

Only $v_f > 0$ is physical, so that $(-)$ is dropped and $|x_i| = x_0$ is used.

- More compression $\leftrightarrow$ faster
- Stiffer spring $\leftrightarrow$ faster
- Larger mass $\leftrightarrow$ slower

Expression matches expectations

$$v_f \propto |x_i| = x_o \checkmark$$

$$v_f \propto \sqrt{k} \checkmark$$

$$v_f \propto \sqrt{1/m} \checkmark$$

Note: Checking qualitative predictions of the final equation is normally only done in one's head; however, it should be done to check if predictions match expectations. If not, then a mistake was made, and this step can help point to the origin of the error.

$$v_f = \left(\sqrt{\frac{k}{m}}\right)(x_o)$$

**Evaluate final expression numerically (6)**

$$= \sqrt{\frac{2.0 \text{ N/m}}{0.25 \text{ kg}}} \cdot 0.20 \text{ m}$$

$$= \left(\sqrt{\frac{2.0}{0.20}} \cdot 0.20\right)\left(\sqrt{\frac{N/m}{kg}} \cdot m\right)$$

$$\boxed{v_f = 0.57 \text{ m/s}}$$

Answer for mass = 0.20 kg

$$v_f = \left(\sqrt{\frac{k}{m}}\right)(x_o)$$

$$= \sqrt{\frac{2.0 \text{ N/m}}{0.50 \text{ kg}}} \cdot 0.20 \text{ m}$$

$$\boxed{v_f = 0.40 \text{ m/s}}$$

Answer for mass = 0.50 kg

# EXAMPLE 4.4

A cart is set rolling up a ramp. The cart has a motor that applies a con-stant force that assists its motion up the ramp. If the initial speed of the cart is 0.50 m/s, the mass of the cart is 0.30 kg, and the slope of the ramp is 10°, for the force having magnitudes of 0.50 N, 0.40 N, 0.30 N, 0.20 N, 0.10 N, and 0.00 N, how far up the ramp will the cart go before coming to rest?

| SOLUTION: | ANNOTATION/COMMENT (STEP #): |
|---|---|
| | Before/After with change in speed and arrangement of system: Energy approach is appropriate to use. |

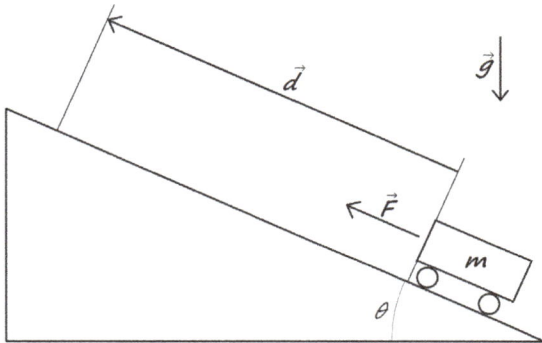

Using diagram to help define key variables and parameters (1)

$m = 0.30$ kg, $\theta = 10°$,

$F = 0.40$ N, $g = 9.8 \dfrac{m}{s^2}$

$v_i = 0.50 \dfrac{m}{s}$, $v_f = 0$, $d = ?$

Define variables being used for later simplification (1)

97

**SOLUTION:**

**ANNOTATION/COMMENT (STEP #):**

$$\begin{pmatrix} 1\ object \\ moving \end{pmatrix} \Leftrightarrow \begin{pmatrix} 1\ K.E.\ term \\ per\ side\ of \\ equation \end{pmatrix}$$

$$\begin{pmatrix} 1\ interaction\ by \\ a\ conservative \\ force\ (gravity) \end{pmatrix} \Leftrightarrow \begin{pmatrix} 1\ P.E.\ term \\ per\ side\ of \\ equation \end{pmatrix}$$

$$\begin{pmatrix} Non- \\ conservative \\ work\ done \\ by\ one\ force \end{pmatrix} \Leftrightarrow (1\ W_{nc}\ term)$$

**Identify the number of each type of term to be used in general expression (1)**

Note: This step is normally done in one's head.

$$E_f = E_i + \sum W_{nc}$$
$$K_f + U_f = K_i + U_i + W_F$$

**Start with fundamental relationship and fill in the appropriate energy terms (2)**

$$W_F = \vec{F} \cdot \vec{d} = +Fd$$

Identify proper form for the work term.
$$\vec{F} \| \vec{d} \therefore \vec{F} \cdot \vec{d} = +Fd$$

$$0 + U_f = K_i + 0 + (+Fd)$$

Define bottom of ramp as $y = 0$ so that $y_i = 0$ and thus $U_i = 0$. Also, enter $K_f = 0$ because the object comes to rest. Explicitly used plus sign in ($+Fd$) to help with checking algebra if needed later.

$$mgy_f = \tfrac{1}{2}mv_i^2 + Fd$$

**Put in specific expressions for each energy term (3)**

$$y_f = d\sin\theta$$

Supplemental equation relating height to distance traveled using geometry

$$mgy_f = \tfrac{1}{2}mv_i^2 + Fd$$
$$mgd\sin\theta = \tfrac{1}{2}mv_i^2 + Fd$$

**Put in variables based on the coordinate system and parameters being used (3)**

$$mgd\sin\theta = \tfrac{1}{2}mv_i^2 + Fd$$
$$mgd\sin\theta - Fd = \tfrac{1}{2}mv_i^2$$
$$d(mg\sin\theta - F) = \tfrac{1}{2}mv_i^2$$

**Do algebra (4)**

**SOLUTION:**

$$d = \frac{mv_i^2}{2(mg\sin\theta - F)}$$

$$[d] = \frac{[mv_i^2]}{[2(mg\sin\theta - F)]}$$

$$= \frac{kg \cdot (m/s)^2}{(1)\left[kg \cdot \dfrac{m}{s^2}\right](1) - N}$$

$$= \frac{kg \cdot (m/s)^2}{\left(kg \cdot \dfrac{m}{s^2} - N\left[\dfrac{kg \cdot (m/s^2)}{N}\right]\right)}$$

$$= \frac{kg \cdot (m^2/s^2)}{kg \cdot (m/s^2)} = \frac{\cancel{kg \cdot (m/s^2)}}{\cancel{kg \cdot (m/s^2)}}\, m$$

$$[d] = m \checkmark$$

$$d = \frac{mv_i^2}{2(mg\sin\theta - F)}$$

$0 \leq F \leq mg\sin\theta$:
denominator smaller → d larger ✓

$F = mg\sin\theta$:
denominator = 0 → d = ∞ ✓

$mg\sin\theta < F$:
denominator < 0 → d < 0 ✓

**ANNOTATION/COMMENT (STEP #):**

Final expression; algebra done.

**Unit analysis (5)**

Note: Performing unit analysis is normally done in one's head. But for complicated expressions, it is worth doing explicitly so that moving to the next step can be justified due to a lack of any noticeable errors.

Units are meters as expected

**Check predictions of final relationship (5)**

Note: This step is normally done "by inspection" (i.e., in one's head). If the expression makes unphysical predictions, a mistake was made or the situation being described is unphysical.

With $F > 0$, the car will go further up track than if $F = 0$.

When $F$ is large enough, the car never slows down.

If $F$ is too large, the solution is outside the limits of applicability because $d$ can only be larger than zero.

| SOLUTION: | ANNOTATION/COMMENT (STEP #): |
|---|---|

**Numerical evaluation of expression (6)**

$$d = \frac{mv_i^2}{2(mg\sin\theta - F)}$$

$$= \frac{(0.30\ kg)(0.50\ m/s)^2}{2(0.30\ kg \cdot 9.8\frac{m}{s^2} \cdot \sin 10° - F(N))}$$

Here used $F(N)$ to designate that this is the value of the force in units of Newtons.

$$d = \left| \frac{0.0375}{(0.5105 - F(N))} \right| m$$

Note: Solving the equations symbolically helps when needing to compute the final expression for many different values.

| F | d |
|---|---|
| 0.50 N | 3.6 m |
| 0.40 N | 0.34 m |
| 0.30 N | 0.18 m |
| 0.20 N | 0.12 m |
| 0.10 N | 0.091 m |
| 0.00 N | 0.073 m |

Trends in the effect of different magnitude of the force are as expected.

Notice that if the applied force $F$ equaled 0.5105 N = $mg \sin(\theta)$, then the denominator would have been zero, and the distance would have been infinite.

# EXAMPLE 4.5

Two blocks are connected by string that goes over a massless pulley. One block hangs vertically and the other slides on a table top. The tabletop is **not** frictionless. Find an expression for how fast the hanging block is traveling after it has fallen a distance $d$ if it is released from rest.

**SOLUTION:**

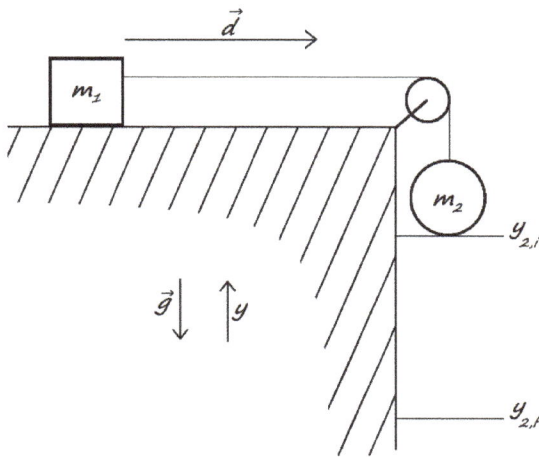

$v_f = ?$

$$\begin{pmatrix} 2 \text{ objects} \\ \text{moving} \end{pmatrix} \Leftrightarrow \begin{pmatrix} 2 \text{ K.E. terms} \\ \text{per side of} \\ \text{equation} \end{pmatrix}$$

$$\begin{pmatrix} 1 \text{ interaction by} \\ \text{a conservative} \\ \text{force (gravity)} \end{pmatrix} \Leftrightarrow \begin{pmatrix} 1 \text{ P.E. term} \\ \text{per side of} \\ \text{equation} \end{pmatrix}$$

$$\begin{pmatrix} \text{Non-} \\ \text{conservative} \\ \text{work done} \\ \text{by one force} \end{pmatrix} \Leftrightarrow \begin{pmatrix} 1 \ W_{nc} \text{ term} \end{pmatrix}$$

**ANNOTATION/COMMENT (STEP #):**

Initial/final states and speeds and forces acting over a distance: Energy approach, including work is appropriate to use.

**Using diagram to help define key variables and parameters (1)**

**Listing variables and data to use (1)**

No values are given for any parameters.

**Identify number of each type of term to be used in general expression (1)**

Note: This step is normally done in one's head.

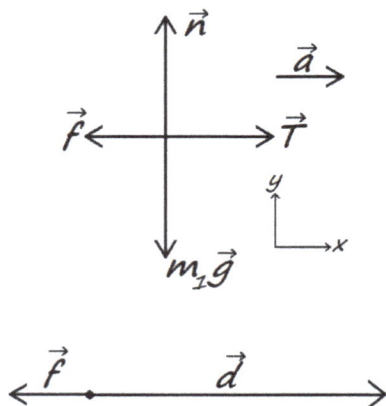

| SOLUTION: | ANNOTATION/COMMENT (STEP #): |
|---|---|
| $$E_f = E_i + \sum W_{nc}$$ $$(K_{1,f} + K_{2,f} + U_{1f}) = (K_{1,i} + K_{2,i} + U_{1,i})$$ $$+ (W_{friction})$$ | **Start with fundamental relationship and fill in the appropriate energy terms (2)** |
| $$\tfrac{1}{2}mv^2_{1,f} + \tfrac{1}{2}mv^2_{2,f} + mgy_{1,f} =$$ $$\tfrac{1}{2}mv^2_{1,i} + \tfrac{1}{2}mv^2_{2,i} + mgy_{1,i} + W_{friction}$$ | **Put in appropriate expressions for each term (3)** |
| Let $y_f = 0$, $y_i - y_f = d$ and because one string, let $$v_{1,f} = v_{2,f} = v_f \text{ and } v_{1,i} = v_{2,i} = v_i = 0$$ | **Define coordinate system so that terms may be set to zero and identify any supplemental expressions (3)** |
| $$\tfrac{1}{2}m_1v^2_f + \tfrac{1}{2}m_2v^2_f + m_2gy_f =$$ $$0 + 0 + m_2gy_i + W_{friction}$$ $$\tfrac{1}{2}(m_1 + m_2)v^2_f = m_2g(y_i - y_f) + W_{friction}$$ $$\tfrac{1}{2}(m_1 + m_2)v^2_f = m_2gd + W_{friction}$$ | **Simplify the main expression using supplemental equations and any special values (3)** |
| $$W_{friction} = \vec{f} \cdot \vec{d} = ?$$ $$|\vec{f}| = f = \mu n$$ | **Create an expression for work due to friction to put into the main expression (3)** |
|  | **Free-body diagram for use with Newton's Second Law to identify the magnitude of the friction force.** |
|  | **Force and displacement vectors to be used in finding expression for the work done by friction.** |

## SOLUTION:

$$\sum \vec{F} = m\vec{a}$$
$$\vec{f} + \vec{n} + m_1\vec{g} + \vec{T} = m_1\vec{a}$$
$$-f\hat{i} + n\hat{j} - m_1 g\hat{j} + T\hat{i} = m_1 a\hat{i} + 0\hat{j}$$

$$x) -f + T = m_1 a$$
$$y) \; n - m_1 g = 0, \; \rightarrow n = m_1 g$$

$$f = \mu n = \mu(n) = \mu m g$$

$$\vec{f} = -\mu mg \, \hat{i}, \; \vec{d} = +d \, \hat{i}$$

$$W_{friction} = \vec{f} \cdot \vec{d}$$
$$= (-\mu m_1 g \, \hat{i}) \cdot (d\hat{i})$$
$$= -\mu m_1 gd(\hat{i} \cdot \hat{i})$$
$$= -\mu m_1 gd(1)$$
$$W_{friction} = -\mu m_1 gd$$

$$\tfrac{1}{2}(m_1 + m_2)v_f^2 = m_2 gd + W_{friction}$$
$$\tfrac{1}{2}(m_1 + m_2)v_f^2 = m_2 gd + (-\mu m_1 gd)$$

$$\tfrac{1}{2}(m_1 + m_2)v_f^2 = m_2 gd - \mu m_1 gd$$
$$v_f^2 = \frac{m_2 gd - \mu m_1 gd}{\tfrac{1}{2}(m_1 + m_2)}$$
$$= 2gd \frac{(m_2 - \mu m_1)}{(m_2 + m_1)}$$

$$\boxed{v_f = (2gd)^{\frac{1}{2}} \left[ \frac{m_2 - \mu m_1}{m_2 + m_1} \right]^{\frac{1}{2}}}$$

## ANNOTATION/COMMENT (STEP #):

Using Newton's Second Law to properly identify the form of the friction force.

Final scalar equations derived from using Newton's Second Law.

Using algebra to solve for $\vec{f}$ and $\vec{d}$ for use in calculating $W_{friction}$.

Evaluating the expression for the work done by friction showing the full mathematical approach to demonstrate how it is done. An alternative would be to write "$\vec{f} \, || -\vec{d}$" and then write $W_{nc} = -fd < 0$, followed by the last line.

**Final substitutions and simplification (3)**

**Physics done; now use algebra (4)**

Expression solved for final speed.

## SOLUTION:

$$[v_f] = \left[ (2gd)^{1/2} \left| \frac{m_2 - \mu m_1}{m_2 + m_1} \right|^{1/2} \right]$$

$$= [2gd]^{1/2} \left| \frac{m_2 - \mu m_1}{m_2 + m_1} \right|^{1/2}$$

$$= \left( 1 \left[ \frac{m}{s^2} \right] m \right)^{1/2} \left( \frac{kg - (1)kg}{kg + kg} \right)^{1/2}$$

$$= \left| \frac{m^2}{s^2} \right|^{1/2} (1)^{1/2}$$

$$[v_f] = \frac{m}{s} \checkmark$$

$$\boxed{\mu \to 0}$$

$$v_f = (2gd)^{1/2} \left| \frac{m_2 - (0)m_1}{m_2 + m_1} \right|^{1/2}$$

$$= (2gd)^{1/2} \left| \frac{m_2}{m_2 + m_1} \right|^{1/2}$$

$$\to v_f(\mu = 0) > v_f(\mu \neq 0) \checkmark$$

$$\boxed{m_1 \to 0}$$

$$v_f = (2gd)^{1/2} \left| \frac{m_2 - \mu(0)}{m_2 + m_1} \right|^{1/2}$$

$$= (2gd)^{1/2} \left| \frac{m_2}{m_2 + (0)} \right|^{1/2}$$

$$v_f = (2gd)^{1/2}$$

$$\to v_f(m_1 = 0) > v_f(m_1 \neq 0) \checkmark$$

## ANNOTATION/COMMENT (STEP #):

### Unit analysis (5)

Note: This step is normally done in one's head, but it is still important to do. If the units are not as expected, then there must be an algebra error. Now is the time to stop, go back, and fix it.

Expect final answer to be in units of meters/second.

### Check predictions of final relationship to expectations and physicalness (5)

Expect the final speed to be faster if no friction.

Note: Doing this step explicitly is optional but should at least be done in one's head as a quick double-check of one's work.

Expect that if $m_1 = 0$, then $m_2$ would be in free-fall and fall faster than if $m_1 \neq 0$.

## SOLUTION:

$$v_f = (2gd)^{1/2}\left[\frac{m_2 - \mu m_1}{m_2 + m_1}\right]^{1/2}$$

For $v_f \in \mathbb{R}$,

$$2gd \overset{!}{\geq} 0 \quad \text{and} \quad \left(\frac{m_2 - \mu m_1}{m_2 + m_1}\right) \overset{!}{\geq} 0$$

and thus

$$m_2 - \mu m_1 \overset{!}{\geq} 0$$

$$m_2 \overset{!}{\geq} \mu m_1$$

$$\therefore \quad \mu \overset{!}{\leq} m_2 / m_1$$

$$v_f = (2gd)^{1/2}\left[\frac{m_2 - \mu m_1}{m_2 + m_1}\right]^{1/2}$$

## ANNOTATION/COMMENT (STEP #):

Checking limits on the final speed based on the result having to be a real number.

For the expression for the speed to be physical, the values in each bracket must be larger than zero so that the overall expression is real.

Thus, if $\mu$ is too large, there can be no solution for $v_f$. Otherwise, it would result in $v_f < 0$, which would correspond to the hanging block going upwards.

This relationship can also be interpreted as consistent with friction coefficients being limited to values between zero and one; $0 \leq \mu \leq 1$ to be physical.

By combining $\mu \overset{!}{\leq} m_2 / m_1$ and $0 \leq \mu \leq 1$, we find that $m_1 \overset{!}{\leq} m_2$ which is what would be expected for the motion to be possible.

> Note: This expression with the exclamation mark over the "$\overset{!}{\leq}$" is read as "must be less than or equal to."

Thus, the final expression makes physical predictions.

Final expression

# 5

# Solving Conservation of Momentum Problems

## TARGETS AND GOALS

**In this chapter, you will learn the following:**

- ✔ **How to identify if the use of Conservation of Momentum is appropriate.**

- ✔ **How to apply it using a general approach that will work for all problems.**

- ✔ **How to test final derived relationships against expected behavior of the system.**

## Introduction

The law of Conservation of Momentum is one of the three conservation laws of mechanics, the other two being Conservation of Energy and Conservation of Angular Momentum. As such, it is characterized as a before/after or initial/final type of problem, with some quantity being the same value for all time.[1] However, unlike energy, momentum (as with angular momentum) is a vector quantity. In other words, not only is the amount of momentum conserved but also its direction. The vector quality of the momentum will be the critical aspect to understand to apply the law of Conservation of Momentum properly.

### Skills and Knowledge Needed to Solve Conservation of Momentum Problems

◊ *Conceptual understanding*

- Knowledge of what momentum is and that it is a vector quantity

- How to apply a conservation law in problem-solving

◊ *Ability to work with vector quantitates*

- How to resolve vectors into component form

◊ *Word problem skills*

- How to re-express physical and mathematical concepts expressed in words as mathematical expressions

- Comfort with solving a system of equations

## Nature of Conservation of Momentum Problems

The law of Conservation of Momentum states that the sum of all the momentum of the particles in a system is constant for all times as long as there are no forces applied from outside the system. This law applies whether or not mechanical energy is conserved or if the particles in the system break apart or combine.

**Law of Conservation of Momentum**

$$\left(\sum_i \vec{p}_i\right) = \vec{P}_{total} = \text{Constant} \quad \text{or} \quad \left(\sum_i \vec{p}_i\right)_{init} = \left(\sum_j \vec{p}_j\right)_{final}$$

*The total momentum of an isolated system defined within an inertial reference frame is constant for all time.*

---

1    If you have wondered if the conservation laws of mechanics are connected, the answer is yes. The derivation that shows this is called "Noether's Theorem" and was published in 1918 by Emmy Noether. It is a standard part of junior-level courses on analytical mechanics. See John Taylor's "Classical Mechanics" and Jerry Marion's and Stephen Thorton's "Classical Dynamics."

Key things to remember when using the law of Conservation of Energy:

1. Independence of the number of objects in the system: The number of objects within a system can change, for example, when two asteroids collide and stick together. Therefore, the number of terms in the sums for the initial and final states can differ.
2. Independence of types of forces: Using the law of Conservation of Momentum does not require knowledge of the nature of the forces of interaction, only that the particles are interacting within an isolated system.
3. Initial/final states are arbitrary: Because for an isolated system, the momentum is constant for all time, we are free to pick the ones that are most convenient.
4. Momentum is a vector quantity: Magnitude and direction both must be included in the problem formulation.

# Caveats about the Use of the Law of Conservation of Momentum

In general, any conservation law can be quite useful. Which one to use will be determined by the information desired about the objects within the system. For example, when only translational speed at some point is desired, a good candidate to use would be the law of Conservation of Energy, where the kinetic energy ($K = \frac{1}{2}mv^2$) is a scalar value and a function of the speed (not velocity). However, if both speed and direction of an object after or during an interaction with another body is desired, then using the law of Conservation of Momentum should be considered.

Another factor in deciding which law to use is whether the force of interaction is known. If not, then Conservation of Momentum is likely a good choice. This is because no matter the type of interaction, the forces exerted on each other by a pair of objects will have the same magnitude and be in opposite directions as described by Newton's Third Law. A consequence is that even interactions that are not elastic, or where mechanical energy is not conserved, can be described by the law of Conservation of Momentum.

## Choosing the Reference Frame

The concept of the center of mass of a system and its motion naturally arises when talking about the motion of a system of particles.

If no external forces are acting on the system, by applying Newton's Second Law, we see that the sum of all of the momenta is constant and thus, so is the velocity of the center of mass.

$$\text{If } \sum \vec{F}_{ext} = 0, \text{ then } \sum \vec{F}_{ext} = \frac{d\vec{P}_{total}}{dt} = \frac{d}{dt}(M_{tot}\vec{V}_{cm}) = 0 \text{ where } \frac{d\vec{V}_{cm}}{dt} = 0 \text{ and thus } \vec{V}_{cm} = \text{constant}$$

Because the motion of the center of mass is constant, we are free to redefine the problem in any other inertial frame by the addition of the same constant velocity to each of the velocities in the expression with the result being that the new expression will still be valid.

In order to transform to the center-of-mass reference frame, we subtract the amount $M_{tot}\vec{V}_{cm} = \left(\sum_i m_i\right)\vec{V}_{cm}$ from both sides of the equation for the total momentum, $\sum_i \vec{p}_i = M_{tot}\vec{V}_{cm}$, which allows us to rewrite the expression for the conservation of momentum in terms of the velocities and momentum measured with respect to the motion of the center of mass, $\vec{v}_i'$ and $\vec{P}'$, where $\vec{v}_i' = \vec{v}_i - \vec{V}_{cm}$:

$$\sum_i \vec{p}_i = M_{tot}\vec{V}_{cm}$$

$$\sum_i (m_i\vec{v}_i) - M_{tot}\vec{V}_{cm} = M_{tot}\vec{V}_{cm} - M_{tot}\vec{V}_{cm}$$

$$\sum_i (m_i\vec{v}_i) - \left(\sum_i m_i\right)\vec{V}_{cm} = 0$$

$$\sum_i (m_i(\vec{v}_i - \vec{V}_{cm})) = 0$$

$$\sum_i \vec{p}_i' = 0$$

For certain problems, solving in the center-of-mass frame can save much work and be more intuitive conceptually. An example of where this might be useful is when we consider the relative motions of the parts of an object that breaks apart, for example, a space probe moving in a straight line that ejects some part of itself into a planet's atmosphere to do measurements.

### Challenges of Doing Conservation of Momentum Problems

The law of Conservation of Momentum is powerful. Part of what makes it powerful is how general it is and how simple the basic mathematical expression is. However, this simplicity and power mean that there is much that can be missed if the mathematical representation is seen as a formula to be used blindly with no underlying understanding. Like with energy, the amount of momentum associated with an object cannot be sensed or measured directly and instead can only be inferred. Because of this, a common mistake is to assume that if momentum is conserved so is the sum of the initial and final velocities.

Another common potential pitfall is due to the vector nature of momentum. When the motion being described has more than one component, the one vector equation must be decomposed into two (or three) scalar equations, one for motion along each axis. This is the same approach that is used when describing the velocity of an object in more than one dimension:

$$\left.\begin{aligned} \vec{v}(t) &= \vec{v}_i + \vec{a}t \\ &= (v_{x,i}\hat{i} + v_{y,i}\hat{j}) + (a_x\hat{i} + a_y\hat{j})t \\ v_x(t)\hat{i} + v_y(t)\hat{j} &= (v_{x,i} + a_x t)\hat{i} + (v_{y,i} + a_y t)\hat{j} \end{aligned}\right\} \Rightarrow \begin{cases} v_x(t) = v_{x,i} + a_x t \\ v_y(t) = v_{y,i} + a_y t \end{cases}$$

Furthermore, keeping track of the signs of the prefactors for the terms expressed using unit-vectors and in the scalar equations is mandatory if errors are to be avoided. This advice applies even for one-dimensional problems where a positive sign (+) will indicate the forward direction and a negative sign (−) will indicate the backward direction.

If in the final analysis of a problem you find that the predicted behavior is the opposite of what you expected, you likely have made a sign error (e.g., a gun at rest is fired, and both the bullet and the gun are predicted to move forward rather than in opposite directions as one would expect).

## The Problem-Solving Steps for Conservation of Momentum Problems

As with all types of physics problems, it all comes down to knowing how to use the tools in your toolbox. But first, one must identify the best tool for the job. To this end, as mentioned elsewhere, "*A well understood problem is a problem half-solved.*" In practice, this consists of learning to see the general characteristics of a problem that you can use to help you identify the most appropriate physics principle to use.

### Identify the Type of Problem

Problems appropriate for using the law of Conservation of Momentum will have several characteristics. An obvious first clue to consider is if the information given about the system involves the mass of the objects and their velocities. Another clue is if there are identifiable initial and final states for the system defined in terms of the velocities or momenta of the objects in the system. Furthermore, if the nature of the forces involved is not identified, then Conservation of Momentum should likely be used.

### Hallmarks of a Conservation of Momentum Problem

- Identifiable initial and final states for a closed system in an inertial reference frame that can be defined in terms of momenta or velocities and masses of the objects

- The nature of the interactions is not provided or not relevant

The first bullet point describes the circumstances and information needed to apply the law of Conservation of Momentum. The second bullet point is a reminder that the details of the nature of the interactions are not necessary to use the law of Conservation of Momentum.

### Recall and Apply the Physics Principle

Although we almost always use the mathematical form in practice, like with all fundamental physics principles of mechanics, the mathematical form must also be understood conceptually. Without the underlying conceptual understanding, any equation is nothing much more than a magical "black-box" formula that can only be used for the simplest of problems. This in contrast to being able to see that the equation represents a very powerful and general concept. With this deeper understanding, the principle becomes easier to use and useful for many more situations than first expected.

- *Mathematical form*: $\left(\sum_i \vec{p}_i\right)_{init} = \left(\sum_j \vec{p}_j\right)_{final}$ or $\vec{P}_{total_{init}} = \vec{P}_{total_{final}}$

- *Conceptual form*: *For an isolated system defined within an inertial reference frame, the sum of the momentum of all of the objects in an initial state will be equal to the sum of the momentum of all of the objects in a final state. This definition does not preclude the objects of the system combining or breaking apart.*

## Applying the Law of Conservation of Momentum

To use the law of Conservation of Momentum, we start with the key equation and fill in each of the terms. For the initial side, we *sum vectorially* all of the momenta for each object in the system and do the same for all of the momenta in the final state. For the case shown below of a system for which there are two objects initially and three later, on the initial side the momentum terms have been labeled 1 and 2, and for the final side, they have been labeled a, b, and c. Because the two momenta in the initial state were in perpendicular directions, a coordinate system in which they were parallel to the x and y directions could be chosen to make expressing as many vectors as possible as simple as possible.

$$\left(\sum_i \vec{p}_i\right)_{init} = \left(\sum_j \vec{p}_j\right)_{final}$$

$$(\vec{p}_1 + \vec{p}_2) = (\vec{p}_a + \vec{p}_b + \vec{p}_c)$$

**Initial**                                        **Final**

In a vector relationship, a picture of the system in which to define the motion is mandatory. It is important not to jump ahead and input negative signs into the vector equation to imply direction. By keeping to vectors and summing, we remind ourselves that it is the sum of these vectors in each state that is conserved. Resolving these into expressions that might have negative signs to denote direction will be done in the next step.

## Simplification of Expressions and Resolving of Vectors into Scalar Equations

In a vector relationship, vector quantities must be resolved into scalar equations so as to be able to use algebra to solve for the parameter of interest. This step will be similar to that used to solve Newton's Second Law problems where the vector form of $\Sigma \vec{F} = m\vec{a}$ is transformed into a set of scalar equations.

The vector quantities are then rearranged in terms of the unit vectors. With the vectors in unit-vector form, the individual terms can be rearranged to collect the terms by direction, making identifying the scalar equations easier.

## Doing the Algebra

This is the step where the scalar equations created from the vector equation are solved for the parameter of interest. Because a system of equations can be the result, being able to solve systems of equations such as 2-equations and 2-unknowns will often be required.

### Re-conceptualizing the Physics

After the algebra is done, then comes the time to check if the answer makes sense. This is where your conceptual understanding of physics and momentum will be used to bolster your confidence in your answers. For instance, a small object interacting with a large object would be expected to have a larger change in speed than the larger object. (This can be inferred from Newton's Third Law.) So, if your final relationship predicts the opposite, that should be interpreted as a clue that perhaps the problem was not set up or solved correctly.

It is also in this step that the limiting cases tests can be performed to check your expectations. These can take the form of asking what the final expression predicts if the angle between the initial trajectories were zero instead of at an angle or if the ratio of the masses was taken to one extreme or another. If the limiting cases, which often are conceptually easier to consider, predict what you would expect, then your expression is likely correct.

Finally, but not necessarily last, would be to check the units of your expressions. If the units for both sides of the equal sign for your expression are not the same, there definitely was an algebra error and your work must be checked.

### Insertion of Numerical Parameters

The final step is to insert the numerical values for the parameters of the system. The answers calculated, of course, must be physical. If, for instance, your final relationship predicts the mass of one of the objects is negative, then there definitely is a problem because that answer is not physical.

# Conservation of Momentum Problem-Solving Steps

Hallmarks: Identifiable initial and final states for a closed system in an inertial reference frame that can be defined in terms of momenta or velocities and masses or moments of inertia of the objects. The nature of the forces might or might not be apparent.

**Law of Conservation of Momentum**

$$\left(\sum_i \vec{p}_i\right)_{init} = \left(\sum_j \vec{p}_j\right)_{final}$$

## Steps

1. **Identify the components of the system\***
   - *Abstract the problem* into figures and mathematical form
     - *Draw a figure* that describes the initial and final states, including vectors for momentum or velocities, and identify the coordinate system to be used
     - *Define the key variables* or relationship between parameters used for the system components
     - *Identify any other relationships* that will or could be used based on the physics (e.g., the Law of Conservation of Energy)

2. **Apply the concept of the Law of Conservation of Momentum\***
   - *Write down the general form for the law* of Conservation of Momentum
   - *For the initial state side of the equation, add vectorially the momentum* of each object in the initial state
   - *For the final state side of the equation, add vectorially the momentum* of each object in the final state

3. **Rewrite the vector equation as a set of scalar equations**
   - *Rewrite vectors in component form and collect terms by unit vector.* For 1-dimensional problems, use + and – signs to designate direction
   - *Identify and write out the scalar equations* associated with each unit vector

4. **Do algebra to solve for the parameter of interest**
   - *Check if the number of equations equals the number of unknowns*
     - *If there are equal numbers* of each, proceed
     - *If not,* identify any additional relationships that might be necessary that were not included in earlier steps
   - *Solve for variable/parameter of interest*

5. **Check "physicalness" of final symbolic relationship**
   - *Perform unit analysis*
   - *Check predictions and special cases* for physicalness

6. **Evaluate expression using data** (if provided)
   - *Check physicalness and reasonableness* of the answer

*\*These are the key steps for this type of problem.*

# EXAMPLE 5.1

A girl on a skateboard is rolling forward at a speed of 3.0 m/s and throws her backpack forward so that its speed measured relative to the ground is 6.0 m/s. After this, her new speed is 2.7 m/s. If the combined mass of the girl and her skateboard is 60 kg, what was the mass of the backpack?

**SOLUTION:**

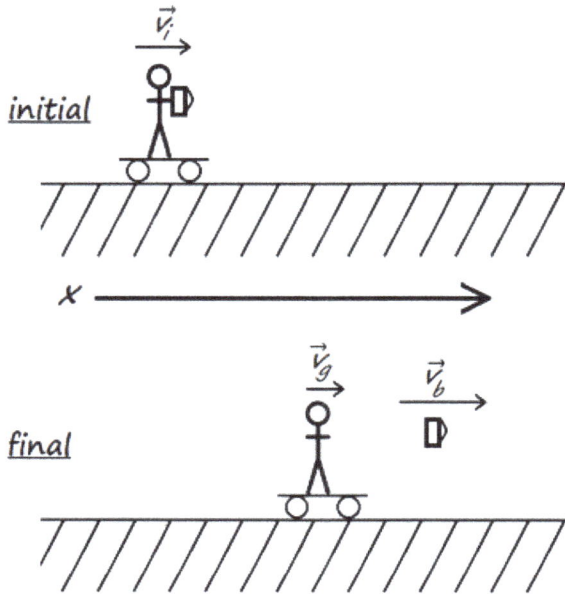

$$m_{girl+skate-board} = m_g = 60 \text{ kg}$$

$$v_i = 3.0 \text{ m/s}, \ v_b = 6.0 \text{ m/s}$$

$$v_g = 2.7 \text{ m/s}$$

$$m_b = ?$$

**ANNOTATION/COMMENT (STEP #):**

Initial and final states defined by changes in velocities and/or momentum: Conservation of Momentum is appropriate.

**Identifying the key parameters of the system (1)**

Drawing figure for initial and final states, including velocity or momentum vectors.

The figure includes the directions of motion and the coordinate system to be used.

Note: Drawing the figure is extremely important so that the relative directions are correctly identified. This should be double-checked against the words of the problem before moving on.

Writing out the data and what is to be found, including supplemental relationships.

Note: For 1-dimensional problems, the signs of the velocity or momentum values denote the direction of motion.

Here all the velocities are defined with respect to the ground (i.e., the "lab reference frame" instead of the "center-of-mass" frame).

## SOLUTION:

$$\left(\sum \vec{p}\right)_i = \left(\sum \vec{p}\right)_f$$

$$\vec{p}_i = \vec{p}_g + \vec{p}_b$$

$$m_i \vec{v}_i = m_g \vec{v}_g + m_b \vec{v}_b$$

$$(m_g + m_b)\vec{v}_i = m_g \vec{v}_g + m_b \vec{v}_b$$

$$(m_g + m_b)v_i = m_g v_g + m_b v_b$$

$$m_g v_i + m_b v_i = m_g v_g + m_b v_b$$

$$m_g v_i - m_g v_g = m_b v_b - m_b v_i$$

$$m_b(v_b - v_i) = m_g(v_i - v_g)$$

$$\boxed{m_b = m_g \frac{(v_i - v_g)}{(v_b - v_i)}}$$

$$[m_b] = \left[m_g \frac{(v_i - v_g)}{(v_b - v_i)}\right] = [m_g]\left[m_g \frac{[v_i - v_g]}{[v_b - v_i]}\right]$$

$$= (kg)\frac{(m/s + m/s)}{(m/s + m/s)} = kg \frac{\cancel{m/s}}{\cancel{m/s}}$$

$$[m_b] = kg \checkmark$$

## ANNOTATION/COMMENT (STEP #):

**Applying the general principle of Conservation of Momentum to the problem (2)**

Writing out this step ensures that the proper number of terms is included on each side of the equation.

This situation is a perfectly inelastic collision, except in reverse.

Note: The terms are only ever summed with no negative signs. The directions of motion will be represented in the plus or minus signs of the component values of the velocities or momenta.

**Rewrite the vector equation as a set of scalar equations (3)**

Note: If a 1-D problem, dropping the vector signs is fine at this step. However, always keep the + signs for the sums. The direction of motion information for each object is represented in the values for their components.

**Do algebra (4)**

Solving for $m_b$. Only four steps were necessary.

The final symbolic expression for $m_b$.

**Checking units (5)**

The velocity terms cancel so that the units are kg as expected.

Note: Checking units should always be done, even if not explicitly written out. This can often be done by inspection (i.e., "in one's head").

| SOLUTION: | ANNOTATION/COMMENT (STEP #): |
|---|---|

**ANNOTATION/COMMENT (STEP #):**

**Checking limiting cases (5)**

$$m_b = m_g \frac{(v_i - v_g)}{(v_b - v_i)}$$

Note: The symbolic equation has been transcribed for convenience.

$(v_i - v_g)$ larger → $m_b$ larger ✓

If the change in the girl's velocity $(v_i - v_g)$ were larger for the same relative velocity of the backpack and the girl $(v_b - v_i)$, then the mass of the backpack would have to be larger to have a significant effect on the girl's motion due to its larger fraction of the total momentum.

$(v_b - v_i)$ larger → $m_b$ smaller ✓

If the difference in the velocity of the backpack and the initial speed were larger for the same change in the girl's velocity $(v_i - v_g)$, then the backpack could not have had much mass; otherwise, it would have caused a greater change in the girl's velocity.

Note: Checking limiting case behavior should always be attempted at least in one's head ("by inspection"), if not written out.

**Evaluating expression for first part of problem (6)**

$$m_b = m_g \frac{(v_i - v_g)}{(v_b - v_i)}$$

Note: The symbolic equation has been transcribed for convenience.

$$m_b = 60\,kg \frac{(3.0\,m/s - 2.7\,m/s)}{(6.0\,m/s - 3.0\,m/s)}$$
$$= 60\,kg(0.1)$$

Substituting the values into the equation to find the mass of the backpack.

$$m_b = 6\,kg$$

The value for the mass of the backpack.

This is not an unreasonable value considering the small change in the girl's velocity and the backpack being thrown at a velocity comparable to that of the girl.

# EXAMPLE 5.2

A physics student is challenged to find the mass of a box sitting in the middle of a frictionless ice-skating rink. Her only tools are an air gun that shoots superballs, a stopwatch, and marks on the ice 1.0 meter apart. She knows that the air gun can shoot the superball with a mass of 50 grams at 20 m/s. When she shoots at the box at an angle so that the ball bounces straight back, she sees that the box takes 5.0 seconds to travel between the marks on the ice. What is the mass of the box? Assume the ice is frictionless and the ball and box have a perfectly elastic collision.

**SOLUTION:**

**ANNOTATION/COMMENT (STEP #):**

Initial and final states defined by changes in velocities and/or momentum: Conservation of Momentum is appropriate.

**Identifying the key parameters of the system (1)**

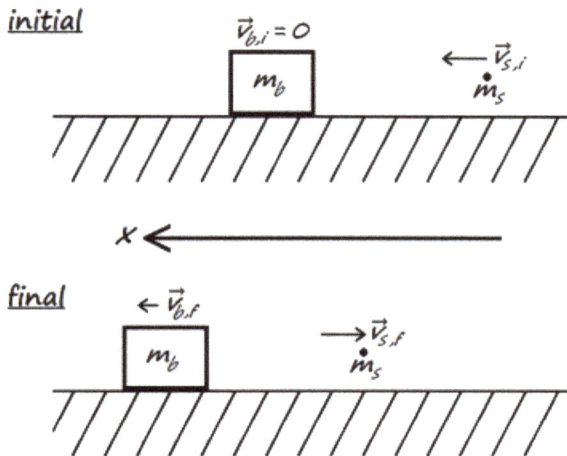

Drawing figure for initial and final states, including velocity or momentum vectors.

The figure includes the directions of motion and the coordinate system to be used.

Note: Drawing the figure is extremely important so that the relative directions are correctly identified. This should be double-checked against the words of the problem before moving on.

$m_s = 50\,g = 0.050\,kg$

Writing out the data supplied and what is to be found, including supplemental any relationships.

$v_{s,i} = +20\,m/s,\ v_{b,i} = 0$

$v_{b,f} = (+1.0\,m/+5.0\,s) = +0.20\,m/s$

$m_b = ?$

**SOLUTION:**

**ANNOTATION/COMMENT (STEP #):**

$$\left(\sum \vec{p}\right)_i = \left(\sum \vec{p}\right)_f$$

$$m_b\vec{v}_{b,i} + m_s\vec{v}_{s,i} = m_b\vec{v}_{b,f} + m_s\vec{v}_{s,f}$$

$$m_b v_{b,i} + m_s v_{s,i} = m_b v_{b,f} + m_s v_{s,f}$$

**Applying the general principle of Conservation of Momentum to the problem (2)**

If we were to use the fundamental form of Conservation of Momentum, the final velocity of the ball, $v_{sf}$, would also be needed.

$$v_{1,f} = \left(\frac{m_1 - m_2}{m_1 + m_2}\right)v_{1,i} + \left(\frac{2m_2}{m_1 + m_2}\right)v_{2,i}$$

$$v_{2,f} = \left(\frac{2m_1}{m_1 + m_2}\right)v_{1,i} + \left(\frac{m_2 - m_1}{m_1 + m_2}\right)v_{2,i}$$

Because this is a 1-dimensional elastic collision, the dedicated equations for this scenario can be used.

Note: The relationships for 1D elastic collisions between two objects come from combining the laws of Conservation of Momentum and Conservation of Energy. With the extra equation used, one less piece of data is required.

Note: Using the 1D elastic collision relationships can save some steps, BUT whether and how to use them requires careful consideration.

$$v_{b,f} = \left(\frac{m_b - m_s}{m_s + m_s}\right)v_{b,i} + \left(\frac{2m_s}{m_b + m_s}\right)v_{s,i}$$

$$v_{s,f} = \left(\frac{2m_b}{m_b + m_s}\right)v_{s,i} + \left(\frac{m_s - m_b}{m_b + m_s}\right)v_{s,i}$$

Assign object 1 to the box (subscript '$b$') and object 2 to the superball (subscript '$s$').

$$v_{b,f} = \left(\frac{m_b - m_s}{m_s + m_s}\right)v_{b,i} + \left(\frac{2m_s}{m_b + m_s}\right)v_{s,i}$$

With only $m_s$, $v_{s,i}$, $v_{b,i}$, $v_{b,f}$ and desiring $m_b$, the equation to use is the first one.

Note: For the special case of elastic 1D collisions, there are no vectors to be converted with directions of motion being represented by the signs of the velocities.

## SOLUTION:

$$v_{b,f} = \left(\frac{m_b - m_s}{m_s + m_s}\right)(0) + \left(\frac{2m_s}{m_b + m_s}\right)v_{s,i}$$

$$\boxed{v_{b,f} = \left(\frac{2m_s}{m_b + m_s}\right)v_{s,i}}$$

$$v_{b,f} = \left(\frac{2m_s}{m_b + m_s}\right)v_{s,i}$$

$$(m_b + m_s)v_{b,f} = 2m_s v_{s,i}$$
$$m_b v_{b,f} + m_s v_{b,f} = 2m_s v_{s,i}$$

$$m_b v_{b,f} = 2m_s v_{s,i} - m_s v_{b,f}$$
$$m_b v_{b,f} = m_s(2v_{s,i} - v_{b,f})$$
$$m_b = m_s\left(\frac{2v_{s,i} - v_{b,f}}{v_{b,f}}\right)$$

$$\boxed{m_b = m_s\left(2\frac{v_{s,i}}{v_{b,f}} - 1\right)}$$

$$[m_b] = \left[m_s\left(2\frac{v_{s,i}}{v_{b,f}} - 1\right)\right]$$

$$= [m_s][2]\left(\frac{v_{s,i}}{v_{b,f}} - [1]\right)$$

$$= (kg)(1)\left(\frac{\cancel{m/s}}{\cancel{m/s}} + (1)\right)$$

$$= kg(1)$$

$$[m_b] = kg \checkmark$$

## ANNOTATION/COMMENT (STEP #):

**Rewrite the vector equation as a set of scalar equations (3)**

When a value is zero, it is appropriate to make that substitution early, as done here.

The final relationship for this problem. (Because it is a 1D problem, there were no vector signs.)

**Do algebra (4)**

Note: The symbolic equation has been transcribed for convenience.

Using algebra to rearrange and solve for $m_b$.

Collected terms by type. Here the velocities are in one factor. This will make checking units and the physicalness of the expression easier.

The final symbolic expression for $m_b$.

**Check Units (5)**

With the similar terms previously grouped together, checking the units is easier.

Note: Units should always be checked to expose and correct errors before moving on. Often this check can be done by inspection without writing out the whole expression.

Units are kg as expected.

| SOLUTION: | ANNOTATION/COMMENT (STEP #): |
|---|---|

**Checking limiting cases (5)**

$$\text{If } \left|\frac{v_{s,i}}{v_{b,f}}\right| \gg 1, \rightarrow (m_b/m_s) \overset{!}{\gg} 1 \quad \checkmark$$

The smaller the box's final speed relative to the ball's initial speed, the larger the box's mass must be compared to the ball. This is expected because if the box had a lot of inertia, it would be hard to accelerate.

$$\text{If } \left|\frac{v_{s,i}}{v_{b,f}}\right| \sim 1, \rightarrow m_b \overset{!}{\sim} m_s$$

This is a special case for when the masses are similar and is the expected behavior. An additional consequence is that the final speed of the ball, $v_{s,f}$, would be zero.

$$\text{If } \left|\frac{v_{s,i}}{v_{b,f}}\right| \sim \frac{1}{2}, \rightarrow m_b \overset{!}{\sim} 0$$

Because $m_b$ must be positive, we know that the fastest the box could go is twice the speed of the ball, but to do so, it would have to have a vanishingly small mass.

Note: Checking the limiting cases is important to do, even if not written out. Doing so can also reveal any limitations on information the relationship can provide or possible outcomes.

Note: The exclamation point over the relational symbol is read as "must."

**Evaluating expression (6)**

$$m_b = m_s\left(2\left|\frac{v_{s,i}}{v_{b,f}}\right| - 1\right)$$

Note: The symbolic equation has been transcribed for convenience.

$$m_b = (0.050\,kg)\left(2\left|\frac{20\,m/s}{0.20\,m/s}\right| - 1\right)$$

Substituting the values into the equation to find the mass of the box.

$$= 9.95\,kg$$

Here, the proper number of significant digits is 2, so the answer is 10 kg.

$$\boxed{m_b = 10\,kg}$$

The value for the mass of the box.

This is a reasonable value considering the relative velocities of the box and ball.

# EXAMPLE 5.3

A 1,000 kg car collides with a second one with a mass of 1,300 kg. The first car has a speed of 12 m/s, and the second is moving towards the first at 2.0 m/s. The final speed of the second one is 8.0 m/s in the op-posite direction of its original motion. A) What is the final velocity of the first car immediately after the collision? B) How much kinetic energy was lost in the collision?

| SOLUTION: | ANNOTATION/COMMENT (STEP #): |
|---|---|
| | Initial and final states defined by changes in velocities and/or momentum: Conservation of Momentum is appropriate. |

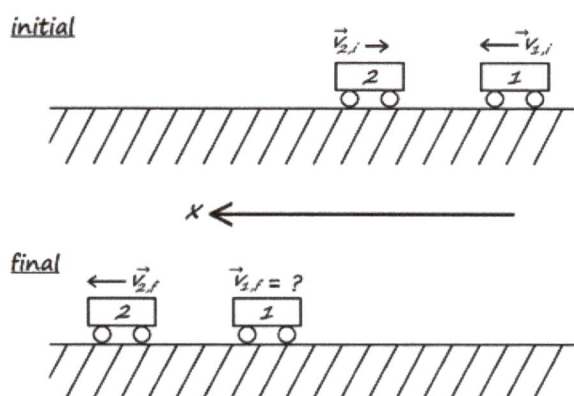

**Identifying the key parameters of the system (1)**

Drawing figure for initial and final states including velocity or momentum vectors.

The figure includes the directions of motion and the coordinate system to be used.

> Note: Drawing the figure is extremely important so that the relative directions are correctly identified. This should be double-checked against the words of the problem before moving on.

$$m_1 = 1,000\,kg,\ m_2 = 1,300\,kg$$

$$v_{1,i} = +12.0\,m/s,\ v_{1,f} = ?$$

$$v_{2,i} = -2.0\,m/s,\ v_{2,f} = +8.0\,m/s$$

$$\Delta K = \left(\sum K\right)_f - \left(\sum K\right)_i = ?$$

Writing out the data and what is to be found, including supplemental relationships.

> Note: For 1-dimensional problems, the signs of the velocity or momentum values denote the direction of motion.

Here all the velocities are defined with respect to the ground, i.e., "the lab reference frame" instead of to the "center of mass frame."

**SOLUTION:**

**ANNOTATION/COMMENT (STEP #):**

**Applying the general principle of Conservation of Momentum to the problem (2)**

Writing out this step ensures that the proper number of terms is included on each side of the equation.

$$\left(\sum \vec{p}\right)_i = \left(\sum \vec{p}\right)_f$$
$$\vec{p}_{1,i} + \vec{p}_{2,i} = \vec{p}_{1,f} + \vec{p}_{2,f}$$

Note: The terms are only ever summed with no negative signs. The directions of motion will be represented in the plus or minus signs of the component values of the velocities or momenta.

**Rewrite the vector equation as a set of scalar equations (3)**

Note: In a 1-D problem, dropping the vector signs is fine at this step. However, always keep the + signs. The direction of motion information for each object is represented in the values for their components.

$$m_1 v_{1,i} + m_2 v_{2,i} = m_1 v_{1,f} + m_2 v_{2,f}$$

**Do algebra (4)**

Solving for $v_{1,f}$. Only two steps were necessary.

$$m_1 v_{1,f} = m_1 v_{1,i} + m_2 v_{2,i} - m_2 v_{2,f}$$

$$\boxed{v_{1,f} = v_{1,i} + \frac{m_2}{m_1}(v_{2,i} - v_{2,f})}$$

The final symbolic expression for $v_{1,f}$.

**Checking units (5)**

$$[v_{1,f}] = \left[v_{1,i} + \frac{m_2}{m_1}(v_{2,i} - v_{2,f})\right]$$

$$= [v_{1,i}] + \left[\frac{m_2}{m_1}\right][(v_{2,i} - v_{2,f})]$$

$$= \left(\frac{m}{s}\right) + \left(\frac{kg}{kg}\right)\left(\frac{m}{s}\right) = \frac{m}{s}$$

The mass terms cancel so that with all of the other terms as velocities, the unit for $v_{1,f}$ is m/s, as expected.

Note: Checking units should always be done, even if not explicitly written out. This can often be done by inspection (i.e., "in one's head").

**SOLUTION:**

**ANNOTATION/COMMENT (STEP #):**

**Checking limiting cases (5)**

$$v_{1,f} = v_{1,i} + \frac{m_2}{m_1}(v_{2,i} - v_{2,f})$$

Note: The symbolic equation has been transcribed for convenience.

for $m_1 >> m_2 \rightarrow v_{1,f} \approx v_{1,i}$ ✓

If $m_2$ were very small compared to $m_1$, only a little of the total momentum would be associated with car 2. Therefore, the momentum of car 1 would not change by much, as predicted.

for $v_{2,i}$ less negative or more positive, then $v_{1,f}$ larger. ✓

If the momentum of car 2 is more to the left (i.e., the + direction), then $v_{1,f}$ would be larger, as predicted.

Note: Checking limiting case behavior should always be attempted at least in one's head, or "by inspection," if not written out.

**Evaluating expression for first part of problem (6)**

$$v_{1,f} = v_{1,i} + \frac{m_2}{m_1}(v_{2,i} - v_{2,f})$$

Note: The symbolic equation has been transcribed for convenience.

$$v_{1,f} = \left(+12.0\frac{m}{s}\right) + \left(\frac{1,300 \text{ kg}}{1,000 \text{ kg}}\right) \times \left(\left(-2.0\frac{m}{s}\right) - \left(8.0\frac{m}{s}\right)\right)$$

$$= 12.0 \text{ m/s} - 13.0 \text{ m/s}$$

$$\boxed{v_{1,f} = -1.0 \text{ m/s}}$$

Substituting the values into the equation for the final velocity of car 1.

Note: In momentum problems, it is easy to forget to put in the correct sign for the velocities.

Final velocity of car 1

Value appears appropriate. Because car 1 and car 2 have similar masses and car 2 has only a small amount of momentum compared to that of car 1, car 1 would be expected to come to rest or only have a small amount of momentum after the collision (or even recoil as predicted here).

**SOLUTION:**

**ANNOTATION/COMMENT (STEP #):**

Now to solve Part B of the problem:

**Applying the change in kinetic energy definition and doing algebra (4)**

$$\Delta K = \left(\sum K\right)_f - \left(\sum K\right)_i$$
$$= (K_{1,f} + K_{2,f}) - (K_{1,i} + K_{2,i})$$
$$= K_{1,f} + K_{2,f} - K_{1,i} - K_{2,i}$$
$$= \tfrac{1}{2}m_1 v_{1,f}^2 + \tfrac{1}{2}m_2 v_{2,f}^2$$
$$\quad - \tfrac{1}{2}m_1 v_{1,i}^2 - \tfrac{1}{2}m_2 v_{2,i}^2$$

Note: All the algebra steps are shown to eliminate ambiguity. In practice, sometimes fewer steps can be used but not so few that the work cannot be checked or full credit awarded.

$$\boxed{\Delta K = \frac{m_1}{2}\left(v_{1,f}^2 - v_{1,i}^2\right) + \frac{m_2}{2}\left(v_{2,f}^2 - v_{2,i}^2\right)}$$

Final symbolic expression for the change in kinetic energy.

**Checking units (5)**

$$[\Delta K] = \left[\frac{m_1}{2}\right]\left[\left(v_{1,f}^2 - v_{1,i}^2\right)\right]$$
$$+ \left[\frac{m_2}{2}\right]\left[\left(v_{2,f}^2 - v_{2,i}^2\right)\right]$$
$$= kg\,\frac{m^2}{s^2} + kg\,\frac{m^2}{s^2} = kg\,\frac{m^2}{s^2}$$
$$= J$$

Note: Checking units is seldom included in written solutions but should always be done by inspection as a quick check. If they are not as expected, then there must have been an error, and therefore, it is not worth continuing until the error is corrected.

Units are Joules, as expected

**Checking limiting cases (5)**

$$\Delta K = \frac{m_1}{2}\left(v_{1,f}^2 - v_{1,i}^2\right) + \frac{m_2}{2}\left(v_{2,f}^2 - v_{2,i}^2\right)$$

For $v_{1,i}^2$ and $v_{2,i}^2$ larger, the more negative $\Delta K$ would be

This predicts that the larger the initial speeds (not velocities) are relative to the final speeds, the more negative the change in kinetic energy of the system will be. This makes sense because the kinetic energy that was lost would have been transformed by energy conservation into work to deform the cars.

Note: Checking the limiting behavior should at least be done by inspection if not written down. It is a quick way to check if the answer makes sense.

**SOLUTION:**

$$\Delta K = \frac{m_1}{2}\left(v^2_{1,f} - v^2_{1,i}\right) + \frac{m_2}{2}\left(v^2_{2,f} - v^2_{2,i}\right)$$

$$\Delta K = \left(\frac{1{,}000\ kg}{2}\right)$$

$$\times \left(\left(-1.0\frac{m}{s}\right)^2 - \left(12.0\frac{m}{s}\right)^2\right)$$

$$+ \left(\frac{1{,}300\ kg}{2}\right)$$

$$\times \left(\left(8.0\frac{m}{s}\right)^2 - \left(-2.0\frac{m}{s}\right)^2\right)$$

$$= 71{,}500\ kg\frac{m}{s} + 39{,}000\ kg\frac{m}{s}$$

$$\boxed{\Delta K = -32.5\ kJ}$$

**ANNOTATION/COMMENT (STEP #):**

**Evaluating expression for the change in kinetic energy (6)**

Note: The symbolic equation has been transcribed for convenience.

Substituting values into the expression for the change in kinetic energy.

Here the negative signs do not affect the answer because the velocities are squared.

Final answer for the change in kinetic energy.

By conservation of energy, $\Delta K \leq 0$ is the only physical answer.

If equal to zero, this would have been an elastic collision.

If less than zero, irreversible work must have been done to deform the cars.

If instead it was found that $\Delta K > 0$, then an error must have been made because this result would be unphysical.

$\Delta K > 0$ is unphysical because the change in potential energy for a collision is generally zero; any change in the total mechanical energy would be in the change in kinetic energy.

In other words, because $E_f = E_i + \sum W_{nc}$, or $\Delta E_{tot} - \sum W_{nc} = 0$, if $\Delta U = 0$, we have $\Delta K_{tot} - \sum W_{nc} = 0$ or $\Delta K_{tot} = \sum W_{nc}$. If negative work is done, then $\Delta K_{tot}$ would have to be less than zero also.

# EXAMPLE 5.4

Two skaters run into each other and move off together. The first is traveling west at 10 m/s, and the second is traveling north at 13 m/s before the collision. The first has a mass of 55 kg and the second a mass of 75 kg. What direction do they move after the collision? What is their final speed?

## SOLUTION:

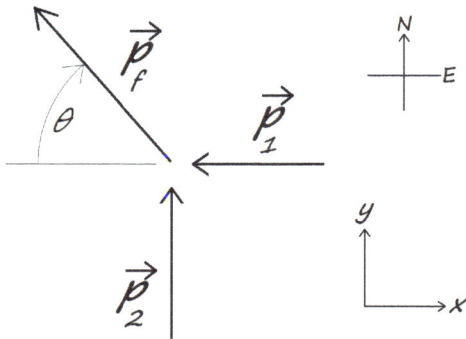

$v_1 = 10\,m/s,\ v_2 = 13\,m/s,$

$m_1 = 55\ kg,\ m_2 = 75\ kg$

$v_f = ?,\ \theta = ?$

$m_f = m_1 + m_2$

$\vec{v}_1 = -v_1\,\hat{i} = -10\,m/s\,\hat{i}$

$\vec{v}_2 = +v_2\,\hat{j} = +13\ m/s\,\hat{j}$

$\vec{v}_f = (-v_{f,x})\hat{i} + (+v_{f,y})\hat{j}$
$\quad = (-v_f\cos\theta)\hat{i} + (v_f\sin\theta)\hat{j}$

## ANNOTATION/COMMENT (STEP #):

Initial and final states defined by changes in velocities and/or momentum: Conservation of Momentum is appropriate.

**Identifying the key parameters of the system (1)**

Drawing figure for initial and final states, including velocity or momentum vectors.

Note: Drawing the figure is extremely important so that the relative directions are correctly identified. This should be double-checked against the words of the problem before moving on.

Used intuition to make an initial guess for the direction of the final momentum vector.

Data values written along with what is to be found.

Supplementary relationship added.

Equations for $m_f$ and vector form of $v_f$ added here.

Note: When possible, it is good practice to use only the magnitudes of the vectors and explicitly use plus and negative signs to represent the direction of the vectors.

| SOLUTION: | ANNOTATION/COMMENT (STEP #): |
|---|---|

**Applying the general principle of Conservation of Momentum to the problem (2)**

$$\left(\sum \vec{p}\right)_i = \left(\sum \vec{p}\right)_f$$

Because this is a perfectly inelastic collision, there are two terms on the initial side and only one on the final side.

$$\vec{p}_1 + \vec{p}_2 = \vec{p}_f$$

For the final side, the mass is the sum of the two skaters, i.e., $m_f = (m_1 + m_2)$.

$$m_1\vec{v}_1 + m_2\vec{v}_2 = m_f\vec{v}_f$$

$$m_1\vec{v}_1 + m_2\vec{v}_2 = (m_1 + m_2)\vec{v}_f$$

Note: Starting with the general expression ensures that the proper number of terms is always included.

Note: The terms are only ever summed with no negative signs. The directions of motion will be represented in the plus or minus signs of the components of the vectors.

**Rewrite vector equation as a set of scalar equations (3)**

$$-m_1 v_1 \hat{i} + m_2 v_2 \hat{j} = (m_1 + m_2) \cdot$$
$$(v_{f,x} \hat{i} + v_{f,y} \hat{j})$$
$$= (m_1 + m_2) \cdot$$
$$(-v_f \cos\theta)\hat{i} + (v_f \sin\theta)\hat{j}$$
$$-m_1 v_1 \hat{i} + m_2 v_2 \hat{j} = -(m_1 + m_2)v_f \cos\theta \hat{i}$$
$$+ (m_1 + m_2)v_f \sin\theta \hat{j}$$

Substituting in the symbolic forms for each term and doing the algebra so that the terms are collected by unit vector.

Final velocity vector expressed in terms of the unit vectors and angle used in the figure: $v_{f,x} = -v_f \cos\theta$ and $v_{f,y} = +v_f \sin\theta$.

$$\boxed{\begin{array}{l} x) \quad m_1 v_1 = (m_1 + m_2)v_f \cos\theta \\ y) \quad m_2 v_2 = (m_1 + m_2)v_f \sin\theta \end{array}}$$

Scalar versions of vector equation.

2 equations with 2 unknowns ($v_f$ and $\theta$); therefore solvable.

Next: eliminate $v_f$ and solve for $\theta$.

| SOLUTION: | ANNOTATION/COMMENT (STEP #): |
|---|---|
| | **Do algebra to solve for the angle (4)** |
| $$\frac{m_1 v_1}{m_2 v_2} = \frac{((m_1 + m_2)v_f \cos\theta)}{((m_1 + m_2)v_f \sin\theta)}$$ | Here, dividing the *x*-equation by the *y*-equation. |
| $$\frac{m_1 v_1}{m_2 v_2} = \frac{(m_1 + m_2)\, v_f}{(m_1 + m_2)\, v_f}\left(\frac{\cos\theta}{\sin\theta}\right)$$ | The final mass and final speed values cancel. |
| $$\frac{m_1 v_1}{m_2 v_2} = \frac{1}{\tan\theta}$$ | |
| $$\boxed{\tan\theta = \frac{m_2 v_2}{m_1 v_1}}$$ | Final expression for the angle |
| | **Checking units of angle relationship (5)** |
| $$[\tan\theta] = \left[\frac{m_2 v_2}{m_1 v_1}\right] = \left(\frac{kg \cdot m/s}{kg \cdot m/s}\right)$$ $$[\tan\theta] = 1 \checkmark$$ | This expression is unitless, as is appropriate for trigonometric functions. Note: Checking units is seldom shown except for complex expressions, but checking, even if only by inspection, is a good way to find any errors. |
| | **Do algebra to solve for the final speed (4)** |
| $$m_1 v_1 = (m_1 + m_2)v_f \cos\theta$$ | Note: The equation has been transcribed for convenience. |
| $$\boxed{v_f = \frac{m_1}{(m_1 + m_2)\cos\theta} v_1}$$ | Expression for final speed |
| | Either the *x* or *y* equation could have been used. Here the *x*-equation was used. In this case, the algebra to solve for $v_f$ only takes one step. |

**SOLUTION:**

$$[v_f] = \left| \frac{m_1}{(m_1 + m_2)\cos\theta} v_1 \right|$$

$$= \left| \frac{m_1}{(m_1 + m_2)} \right| \left| \frac{1}{\cos\theta} \right| [v_1]$$

$$= \left( \frac{\cancel{kg}}{\cancel{kg} + \cancel{kg}} \right) \left( \frac{1}{1} \right) \left( \frac{m}{s} \right)$$

$$[v_f] = \left( \frac{m}{s} \right) \checkmark$$

$$\tan\theta = \frac{m_2 v_2}{m_1 v_1}$$

$$\tan\theta = \frac{(75\,kg)(13\,m/s)}{(55\,kg)(10\,m/s)}$$

$$= \left( \frac{975}{550} \right) \left( \frac{\cancel{kg\cdot m/s}}{\cancel{kg\cdot m/s}} \right) = 1.773$$

$$\theta = \tan^{-1}(1.773)$$

$$\boxed{\theta = 61°}$$

clockwise from negative x-axis

**ANNOTATION/COMMENT (STEP #):**

**Checking units (5)**

Note: Checking units is seldom shown except for complex expressions, but doing the check, even if only by inspection, is a good way to check for errors.

The units are m/s as expected.

**Evaluating expression for the angle that indicates the final direction of motion (6)**

Note: The equation has been transcribed for convenience.

Substitutions for values provided to evaluate expression for the angle.

Final answer for the angle, including how the angle is defined.

The angle is 61 degrees, *clockwise* from the *negative x*-axis. This makes sense since the total momentum in the *y*-direction is greater than the total momentum in the *x*-direction.

Note: If the initial guess for the direction of the final velocity vector had been wrong (e.g., it had both positive *x* and *y* components), then the answers would not have made sense and would indicate that the original setup was likely incorrect.

## SOLUTION:

$$v_f = \frac{m_1}{(m_1 + m_2)\cos\theta} v_1$$

$$v_f = \frac{55 \text{ kg}}{(55 \text{ kg} + 75 \text{ kg})\cos 61°}\left(10\,\frac{m}{s}\right)$$

$$\boxed{v_f = 8.7\,\frac{m}{s}}$$

## ANNOTATION/COMMENT (STEP #):

### Evaluating the value of the final speed (6)

<u>Note:</u> The equation has been transcribed for convenience.

Substitutions for values provided and angle to evaluate expression for $v_f$.

<u>Value for final speed of the skaters.</u>

The answer makes sense because it is comparable to the values of the speeds of the two skaters, who are not too different in their masses.

<u>Note:</u> Because the vectors were defined in terms of their magnitudes with any plus and minus signs used separately to define the direction of a component, any solution for a vector magnitude would be expected to be positive. If a negative value were found, then there was likely an error in the solution.

# EXAMPLE 5.5

A hockey puck sliding in a straight line on ice at 3.5 m/s explodes into three pieces. Each piece moves as follows:

1. $(m_1 = 40$ g) at 30° on left from straight back with $v_1 = 2.0$ m/s.

2. $(m_2 = 75$ g) at 10° to the left from straight forward.

3. $(m_3 = 45$ g) at 45° to the right from straight forward.

What is the speed of the second piece?

| SOLUTION: | ANNOTATION/COMMENT (STEP #): |
|---|---|
| | Initial and final states defined by changes in velocities and/or momentum: Conservation of Momentum is appropriate. |
| | **Identifying the key parameters of the system (1)** |

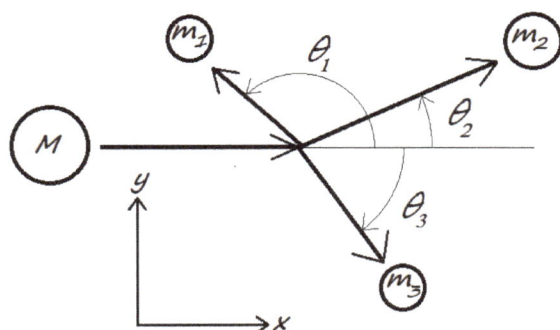

Drawing figure for initial and final states including velocity or momentum vectors.

> Note: Drawing the figure is extremely important so that the relative directions are correctly identified. This should be double-checked against the words of the problem before moving on.

$v_i = 3.5$ m/s, $v_1 = 2.0$ m/s,

$\theta_1 = 150°, \theta_2 = 10°, \theta_3 = -45°$

Direction angles translated into polar coordinates with proper signs.

$m_1 = 40$ g, $m_2 = 75$ g, $m_3 = 45$ g

Data values written along with what is to be found.

$v_2 = ?$

Whether enough information has been provided will be checked later.

$M = m_1 + m_2 + m_3$

Supplementary relationship added.

The extra relationship is for the conservation of mass.

| SOLUTION: | ANNOTATION/COMMENT (STEP #): |
|---|---|

$$\left(\sum \vec{p}\right)_i = \left(\sum \vec{p}\right)_f$$
$$\vec{p}_i = \vec{p}_1 + \vec{p}_2 + \vec{p}_3$$
$$M\vec{v}_i = m_1\vec{v}_1 + m_2\vec{v}_2 + m_3\vec{v}_3$$

**Applying the general principle of Conservation of Momentum to the problem (2)**

This is a perfectly inelastic collision, hence the different number of terms on each side.

$$M v_i \hat{i} = m_1(v_1 \cos\theta_1 \hat{i} + v_1 \sin\theta_1 \hat{j})$$
$$+ m_2(v_2 \cos\theta_2 \hat{i} + v_2 \sin\theta_2 \hat{j})$$
$$+ m_3(v_3 \cos\theta_3 \hat{i} + v_3 \sin\theta_3 \hat{j})$$

**Rewriting the vector equations into a set of scalar equations (3)**

The vectors in unit-vector form have the same structure because all the angles are measured from the same reference point.

$$\left. \begin{matrix} M v_i \hat{i} \\ \\ + 0\hat{j} \end{matrix} \right\} = \begin{cases} \hat{i}(m_1 v_1 \cos\theta_1 + m_2 v_2 \cos\theta_2 \\ \quad + m_3 v_3 \cos\theta_3) \\ + \hat{j}(m_1 v_1 \sin\theta_1 + m_2 v_2 \sin\theta_2 \\ \quad + m_3 v_3 \sin\theta_3) \end{cases}$$

Collecting terms by unit-vector.

The zero-vector for the y-component is added as a placeholder.

x) $M v_i = m_1 v_1 \cos\theta_1 + m_2 v_2 \cos\theta_2$
$\qquad + m_3 v_3 \cos\theta_3$

y) $\quad 0 = m_1 v_1 \sin\theta_1 + m_2 v_2 \sin\theta_2$
$\qquad + m_3 v_3 \sin\theta_3$

The vector equations in scalar form

Knowns:

$m_1, m_2, m_3, \theta_1, \theta_2, \theta_3, v_i, v_1$

Unknowns:

$M, v_2, v_3$

3 unknowns and 3 equations (two from momentum and one from conservation of mass); therefore, the system is solvable.

Note: Checking that the number of equations equals the number of unknowns early will give you confidence that you have started correctly.

$$\boxed{\begin{matrix} M v_i = m_1 v_1 \cos\theta_1 + m_2 v_2 \cos\theta_2 \\ \qquad + m_3 v_3 \cos\theta_3 \\ 0 = m_1 v_1 \sin\theta_1 + m_2 v_2 \sin\theta_2 \\ \qquad + m_3 v_3 \sin\theta_3 \\ M = m_1 + m_2 + m_3 \end{matrix}}$$

Set of three relationships that will be used.

Note: Although the problem is only asking for one of the unknown values, in principle, the others could also be found.

**SOLUTION:**

**ANNOTATION/COMMENT (STEP #):**

**Doing algebra (4)**

$$0 = m_1 v_1 \sin \theta_1 \\ + m_2 v_2 \sin \theta_2 \\ + m_3 v_3 \sin \theta_3$$

Starting with the y-equation so as to be able to eliminate $m_3 v_3$ in the x-equation.

$$m_3 v_3 \sin \theta_3 = -m_1 v_1 \sin \theta_1 \\ - m_2 v_2 \sin \theta_2$$

$$m_3 v_3 = -m_1 v_1 \frac{\sin \theta_1}{\sin \theta_3} \\ - m_2 v_2 \frac{\sin \theta_2}{\sin \theta_3}$$

Expression for y-component motion solved for $m_3 v_3$ so that it can be eliminated. This is necessary because the value for $v_3$ is not given.

$$M v_i = m_1 v_1 \cos \theta_1 + m_2 v_2 \cos \theta_2 \\ + m_3 v_3 \cos \theta_3$$

Expression for x-component motion into which the expression for $m_3 v_3$ will be substituted.

$$= m_1 v_1 \cos \theta_1 + m_2 v_2 \cos \theta_2 \\ + \left( -m_1 v_1 \frac{\sin \theta_1}{\sin \theta_3} \right. \\ \left. -m_2 v_2 \frac{\sin \theta_2}{\sin \theta_3} \right) \cdot \cos \theta_3$$

Substituting the expression for $m_3 v_3$ into x-equation in the parentheses.

Note: If each algebra step is done correctly, the order they are done will not matter because there can only be one possible answer.

$$M v_i = m_1 v_1 \cos \theta_1 - m_1 v_1 \frac{\sin \theta_1}{\tan \theta_3} \\ + m_2 v_2 \cos \theta_2 - m_2 v_2 \frac{\sin \theta_2}{\tan \theta_3}$$

The only unknown value left in the expression is $v_2$.

$$M v_i = m_1 v_1 \left( \cos \theta_1 - \frac{\sin \theta_1}{\tan \theta_3} \right) \\ + m_2 v_2 \left( \cos \theta_2 - \frac{\sin \theta_2}{\tan \theta_3} \right)$$

Because it is an easily calculable value, M is left in the expression.

| SOLUTION: | ANNOTATION/COMMENT (STEP #): |
|---|---|

$$m_2 v_2 \left( \cos\theta_2 - \frac{\sin\theta_2}{\tan\theta_3} \right) =$$

$$M v_i - m_1 v_1 \left( \cos\theta_1 - \frac{\sin\theta_1}{\tan\theta_3} \right)$$

Rearranging to solve for $v_2$ on the left side.

$$\boxed{v_2 = \left[ \frac{M}{m_2} v_i - \frac{m_1}{m_2} v_1 \left( \cos\theta_1 - \frac{\sin\theta_1}{\tan\theta_3} \right) \right] \cdot \left( \cos\theta_2 - \frac{\sin\theta_2}{\tan\theta_3} \right)^{-1}}$$

Final expression for speed of second piece

$$[v_2] = \left[ \left( \frac{M}{m_2} v_i \right. \right.$$

$$\left. - \frac{m_1}{m_2} v_1 \left( \cos\theta_1 - \frac{\sin\theta_1}{\tan\theta_3} \right) \right)$$

$$\left. \cdot \left( \cos\theta_2 - \frac{\sin\theta_2}{\tan\theta_3} \right)^{-1} \right]$$

### Checking units (5)

Note: Although not always explicitly shown, checking units should always be done, even if only "by inspection." It is well worth doing for complicated expressions because it can show if any errors have been made in the previous steps. If no errors are found, then the algebra was likely done correctly.

Note: Collecting like terms in ratios makes checking the units much easier.

$$= \left[ \left[ \frac{M}{m_2} \right] [v] \right.$$

$$\left. + \left[ \frac{m_1}{m_2} \right] [v_1] \left( \cos\theta_1 - \frac{\sin\theta_1}{\tan\theta_3} \right) \right]$$

$$\left. \cdot \left[ \cos\theta_2 - \frac{\sin\theta_2}{\tan\theta_3} \right]^{-1} \right.$$

$$= \left( 1 \cdot \left( \frac{m}{s} \right) + 1 \cdot \left( \frac{m}{s} \right) \cdot 1 \right) \cdot (1+1)^{-1}$$

$$[v_2] = \frac{m}{s} \checkmark$$

Units are m/s as expected.

## SOLUTION:

$$v_2 = \left[ \frac{M}{m_2} v_i - \frac{m_1}{m_2} v_1 \left( \cos\theta_1 - \frac{\sin\theta_1}{\tan\theta_3} \right) \right]$$
$$\cdot \left( \cos\theta_2 - \frac{\sin\theta_2}{\tan\theta_3} \right)^{-1}$$

$$v_2 \sim (v_i - v_1) \checkmark$$

$$v_2 \sim 1/m_2 \checkmark$$

## ANNOTATION/COMMENT (STEP #):

### Checking expectations (5)

Note: Final symbolic expression transcribed for convenience.

This is a complicated expression; therefore, whether it can be analyzed easily is not obvious. This is ok, as we will do the best we can and then compare the final numerical result to our expectations.

We would expect the speed of the second piece to be related to the difference in the speeds of the other two final pieces because one is going forward and the other backward. We see this in the (approximate) difference between the $v_i$ and $v_1$ terms. (The $v_1$ term contains the information about piece 3.)

We would expect the speed of the second piece to be smaller if it had a larger mass. This is because it is momentum, not speed (or velocity), that is conserved. If the mass of piece 2 were large, it would not have to be going very fast to have the same momentum as if it had less mass.

## SOLUTION:

$$V_2 = \left[ \frac{M}{m_2} V_i - \frac{m_1}{m_2} V_1 \left( \cos\theta_1 - \frac{\sin\theta_1}{\tan\theta_3} \right) \right]$$
$$\cdot \left( \cos\theta_2 - \frac{\sin\theta_2}{\tan\theta_3} \right)^{-1}$$

$$V_2 = \left[ \left( \frac{(40 + 75 + 45)g}{75g} \right) \left( 3.5\frac{m}{s} \right) \right.$$
$$- \left( \frac{40g}{75g} \right) \left( 2.0\frac{m}{s} \right)$$
$$\left. \cdot \left( \cos 150^\circ - \frac{\sin 150^\circ}{\tan(-45^\circ)} \right) \right]$$
$$\cdot \left( \cos 10^\circ - \frac{\sin 10^\circ}{\tan(-45^\circ)} \right)^{-1}$$

$$= \left[ 7.47\frac{m}{s} - \left( 1.07\frac{m}{s} \right) \left( -0.866 - \frac{0.500}{-1.00} \right) \right]$$
$$\cdot \left( 0.985 - \frac{0.174}{-1.00} \right)^{-1}$$

$$= \left( 7.46\frac{m}{s} + 0.392\frac{m}{s} \right) (1.15)^{-1}$$

$$\boxed{V_2 = 6.8\frac{m}{s}}$$

## ANNOTATION/COMMENT (STEP #):

**Evaluating expression numerically (6)**

Note: Final symbolic expression transcribed for convenience.

Substituting in the provided data values.

Note: For the purposes of calculation, one more significant digit is kept for the intermediate values. For the final answer, the correct number of digits is used.

Note: It is good practice to include leading zeros before decimal points. This prevents errors in later transcriptions and calculations.

Value for the speed $v_2$

The value is not unreasonable, as the other two pieces have somewhat opposing motion, and piece 2 is in the forward direction with about half of the total mass.

The sign of the value is positive, which indicates that the setup for the vectors was correct. If it had been negative, that would have indicated that the representation in the problem setup of the vector for the second piece was pointing in the wrong (opposite) direction.

# Solving Rotational Dynamics and Angular Momentum Problems

# 6

## Introduction

Rotational dynamics problems can be tricky for some students, but this does not have to be. The notation looks different, and the ability to intuit concepts like torque and momenta of inertia may not come naturally. This is understandable; humans move along straight lines and curved paths. Translation through space is what we know. What we do not normally experience in our day-to-day lives is spinning on an axis and rotation of our bodies.[1]

To bridge this gap between the translational and rotational takes a bit more abstraction. However, like with many other concepts, a bit of guidance in the abstraction of the basic concepts from what we are familiar to those with which we are not so familiar, understanding of those newer concepts will be created with a new set of powerful problem-solving tools becoming available.

### Skills and Knowledge Needed to Solve Rotational Dynamics Problems

◊ *Conceptual understanding*:

- How to work with both translational and rotational physics concepts

- The connections, conceptually and mathematically, between the translational and rotational forms of the main concepts of mechanics

◊ *Familiarity with concepts of rotational dynamics*:

- Rotational kinematics

- Moment of inertia (rotational inertia)

- Torque (rotational force)

- Angular momentum (rotational momentum)

◊ *Word problem skills*:

- How to re-express physical and mathematical concepts expressed in words as mathematical expressions

- Comfort with solving systems of equations

## TARGETS AND GOALS

In this chapter, you will learn the following:

✔ **How to identify when the rotational versions of Newton's Second Law, the Work-Kinetic Energy Theorem, Conservation of Energy, or Conservation of Angular Momentum is appropriate to use.**

✔ **How to apply the rotational versions of Newton's Second Law, the Work-Kinetic Energy Theorem, Conservation of Energy, and Conservation of Angular Momentum, and use them to set up the mathematical expressions that will describe the behavior of a system.**

## The Nature of Rotational Dynamics Problems

Solving rotational dynamics problems requires understanding the rotational versions of the general principles of physics that most students learn first: Kinematics, Newton's Second Law, the Work-Kinetic Energy Theorem, Conservation of Energy,

---

1  There are exceptions. Perhaps one of the more obvious ones would be figure skaters who do spins in their competitions. Another example would be those who cannot help themselves and enjoy spinning in office chairs.

and Conservation of Momentum. The goal should be to see problems of rotational motion and interactions not as some special case but rather as another flavor of dynamics.

Throughout this chapter, the words translational and rotational will often be used to distinguish between the different forms of the main physics concepts and principles of mechanics. This is to remind the reader of the underlying similarity of the two versions of the concepts and that one is not a special case of the other.

To this end, this is a table of the terms that will be used with their translational and rotational form along with their traditional names.

**TABLE 6.1** Traditional Names, Translational Forms, and Rotational Forms

| Physics Concept | Traditional name<br>*Translational form* | Traditional name<br>*Rotational form* |
| --- | --- | --- |
| Inertia | Mass<br>*Translational Inertia* | Moment of Inertia<br>*Rotational Inertia* |
| Force | Force<br>*Translational Force* | Torque<br>*Rotational Force* |
| Momentum | Momentum<br>*Translational Momentum* | Angular Momentum<br>*Rotational Momentum* |
| Kinetic Energy | Kinetic Energy<br>*Translational Kinetic Energy* | Rotational Kinetic Energy<br>*Rotational Kinetic Energy* |
| Work | Work<br>*Translational Work* | Rotational Work<br>*Rotational Work* |
| Displacement | Displacement<br>*Translational Displacement* | Rotation<br>*Rotational Displacement* |
| Velocity | Velocity<br>*Translational Velocity* | Angular Velocity<br>*Rotational Velocity* |
| Acceleration | Acceleration<br>*Translational Acceleration* | Angular Acceleration<br>*Rotational Acceleration* |

# Caveats and Subtilties of Rotational Dynamics Problems

The following discussions of the rotational versions of kinematics, inertia, Newton's Second Law, momentum, kinetic energy, and work are intended as a review of the discussions normally provided in standard textbooks but in the context of problem-solving.

### Rotational Kinematic Quantities

The dynamics of rotating bodies has a fundamental assumption: that the body is rigid. This means that the amount and rate of rotation of a body in terms of the amount of revolution or revolutions will be the same for every part of that body. As such, only one parameter is needed each to specify an amount of rotation, speed of rotation, and acceleration of the rotational rate. For most if not all first-year physics, the mathematical form of rotational kinematics problems will be similar to 1D translational kinematics problems.

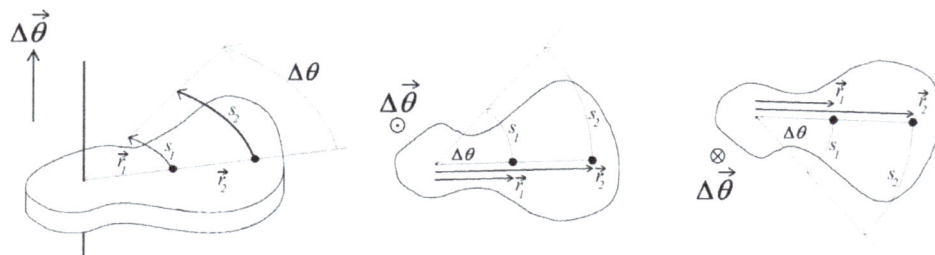

The figure above shows how to represent the rotational displacement vector $\Delta\vec{\theta}$. On the left is the oblique view, with the center and right being the views from above and below, respectively. For each case, the magnitude of the displacement, $\left|\Delta\vec{\theta}\right| = \Delta\theta$, is the same and is positive. We have also used the notation that a vector pointing out of the page is given by the symbol ⊙, a circle with a dot, and a vector pointing into the page is given by ⊗, a circle with a cross in it.[2]

Also identified are two points in the object located at $\vec{r}_1$ and $\vec{r}_2$ relative to the rotation axis and the corresponding lengths of the arcs their motion trace when the object is displaced by $\Delta\vec{\theta}$. The length of these arcs given by $s_1 = |\vec{r}_1||\Delta\vec{\theta}| = r_1(\Delta\theta)$ and $s_2 = |\vec{r}_2||\Delta\vec{\theta}| = r_2(\Delta\theta)$.

These angular displacements and subsequent velocities and accelerations *about* an axis are easily related to the distances, speeds, and accelerations *along* circular paths with the application of some basic math:

$$s = r\theta \Rightarrow \frac{ds}{dt} = \frac{d}{dt}(r\theta) = r\frac{d\theta}{dt} \Rightarrow v_t = r\omega \qquad v_t = r\omega \Rightarrow \frac{dv_t}{dt} = \frac{d}{dt}(r\omega) = r\frac{d\omega}{dt} \Rightarrow a_t = r\alpha$$

A common challenge when solving problems is that we have several accelerations to quantify: the angular acceleration $\alpha$ about an axis, the tangential acceleration $a_t$ along a circular path, and the centripetal acceleration $a_c$ pointing inward, perpendicular to the path. The second and third of these are related to the angular acceleration through these:

$$a_t = r\alpha \text{ and } a_c = \frac{v_t^2}{r} = \frac{(r\omega)^2}{r} = r\omega^2 .$$

## Rotational Units

In order to perform unit analysis in rotational problems, the units of rotational variables must be understood. Although using the units of degrees may seem natural, the preferred unit for measuring rotation is the radian.

Whereas there are 360 degrees in a circle, the definition of radians is that there are $2\pi$ radians in a complete circle (i.e., $2\pi$ radians = 360°). Radians can be considered a *natural unit* because it directly relates an arc length and the angle it subtends, $s = r\theta$, to the circumference of a circle, $c = 2\pi r$, through a simple proportionality. This can be seen when we divide the arc length by the circumference; we find that the ratio of $s/c$ is the same as the ratio of the angle $\theta$ to $2\pi$.

$$\frac{s}{c} = \frac{(\theta r)}{(2\pi r)} = \frac{\theta}{2\pi}\left(\frac{\cancel{r}}{\cancel{r}}\right) = \frac{\theta}{2\pi} .$$

A consequence of this is that for the purposes of unit analysis, <u>radians are unitless</u>, which can be seen in the defining equation for the arc length:

$$s = r\theta \Rightarrow [s] = [r\theta] = [r][\theta] \Rightarrow [\theta] = [s]/[r] = L/L = 1$$

Similarly, we can deduce the units for angular velocity and angular acceleration from the relationships between them and tangential speed and acceleration, respectively:

$$[\omega] = [\text{rad/s}] = 1/\text{T} \qquad \text{and} \qquad [\alpha] = [\text{rad/s}^2] = 1/\text{T}^2$$

In practice, verbally and written, we will use the word "radians," knowing that for unit analysis purposes, radians are unitless. For example:

- $\omega = 2.5/\text{s} = 2.5 \text{ rad/s}$         = "the angular speed is 2.5 radians per second"

- $\alpha = 0.4/\text{s}^2 = 0.4 \text{ rad/s}^2$         = "the angular acceleration is 0.4 radians per second-squared"

Another common conversion that will show up in many physics problems and practical examples is when the rotational speed is given in revolutions per minute or r.p.m. To convert such expressions, we use the identity that 1 revolution = $2\pi$ radians:

$$\omega = 20 \text{ r.p.m.} = 20\frac{\text{rev.}}{\text{min.}} = 20\left(\frac{\cancel{\text{rev.}}}{\cancel{\text{min.}}}\right)\left(\frac{2\pi \text{ rad}}{\cancel{\text{rev.}}}\right)\left(\frac{1 \cancel{\text{min.}}}{60 \text{ s}}\right) = 2.1\frac{\text{rad}}{\text{s}} = \frac{2.1}{\text{s}}$$

In words, "20 r.p.m. equals 2.1 radians per second." It is acceptable to also express it as $\omega = 2.1$ rad/s as long as for the purposes of calculation, it is remembered that radians are unitless.

---

2  These two symbols are often remembered as representing the tip of an arrow pointing toward the reader ⊙, and ⊗ representing the fletching of an arrow pointing away from the reader.

## Rotational Kinematics

As a review, the following is a listing of the standard 1D kinematics equations along with their rotational equivalents. Note the parallel structure of the translational and rotational analogs.

**TABLE 6.2** 1D Kinematics Equations and Relational Equivalents

| 1-Dimensional Translation | Rotation |
|---|---|
| $v_f = v_i + at$ | $\omega_f = \omega_i + \alpha t$ |
| $x_f = x_i + v_i t + \frac{1}{2}at^2$ | $\theta_f = \theta_i + \omega_i t + \frac{1}{2}\alpha t^2$ |
| $v_f^2 - v_i^2 = 2a(x_f - x_i)$ | $\omega_f^2 - \omega_i^2 = 2\alpha(\theta_f - \theta_i)$ |
| $x_f = x_i + v_f t - \frac{1}{2}at^2$ | $\theta_f = \theta_i + \omega_f t - \frac{1}{2}\alpha t^2$ |
| $x_f = x_i + \frac{1}{2}(v_i + v_f)t$ | $\theta_f = \theta_i + \frac{1}{2}(\omega_i + \omega_f)t$ |

## Rotational Inertia; The Moment of Inertia

For objects moving along a path, under translation, the ease of getting that object to change its speed is directly related to how much of it there is: its mass. This is easily seen by a simple rearrangement of Newton's Second Law: $\vec{a} = \vec{F}/m$, where for a given force, the acceleration is inversely proportional to the amount of mass the object possesses.

In contrast, the equivalent measure of how hard it is to get something to change its rotational speed depends on the mass of the object <u>and</u> how that mass is distributed within the body, with the rotational inertia being greater the closer the mass is to the outside boundary of the object.

For simple objects with rotational symmetry that can be defined by an outer radius $R$ and a total mass $M$, the moment of inertia will have the form $I = \beta MR^2$, where the pre-factor $\beta$ is between zero and one and the larger value of $\beta$ corresponds to the mass distribution being more towards the outer edge.

The corresponding rotational version of Newton's Second Law solved for the acceleration and expressed in terms of the parameterized moment of inertia has a similar form to the version for translational motion where the rotational acceleration is inversely proportional to <u>both</u> the total mass and how the mass is distributed as represented by the variable $\beta$: $\vec{\alpha} = \vec{\Gamma}/I = \vec{\Gamma}/\beta MR^2$.

To help develop an intuition for the relationship between the mass distribution and the moment of inertia, table 6.3 lists from left to right objects with larger to smaller moments of inertia that otherwise have the same mass and outside radius.

**TABLE 6.3** Objects with Larger to Smaller Moments of Inertia that Have the Same Mass and Outside Radius

| Shape | Hollow Cylinder or Hoop | Hollow Sphere | Solid Cylinder or Disk | Solid Sphere | Solid Cone |
|---|---|---|---|---|---|
| Moment of Inertia | $I = \frac{1}{1}MR^2$ | $I = \frac{2}{3}MR^2$ | $I = \frac{1}{2}MR^2$ | $I = \frac{2}{5}MR^2$ | $I = \frac{3}{10}MR^2$ |
| Image | | | | | |
| Mass Distribution | farthest from axis | | | | closer to axis |

## Moments of Inertia of Compound Shapes

When an object under translation collides and combines with another by a perfectly inelastic collision, the inertia of the new object is simply the sum of the masses of the two objects. For rotational systems in which two objects collide and stick together, a new moment of inertia must be defined with respect to a possibly new axis of rotation. How is this handled in rotational problems?

Because the moment of inertia of a rigid body is a sum over the moments of inertia of the individual parts (expressed as an integral), we are permitted to break up the limits of integration in any way we like. When the combined object is made up of two familiar objects that share a common rotation axis, we can make the two parts of our integral correspond to the two familiar parts and simply add their respective moments of inertia. This is shown mathematically in the following equation.

$$I_{total} = \int\limits_{\substack{\text{part A}\\+\\\text{part B}}} r^2 \, dm = \left(\int\limits_{\text{part A}} r^2 \, dm\right) + \left(\int\limits_{\text{part B}} r^2 \, dm\right) = I_A + I_B$$

For the case of a hollow cylinder attached to a solid sphere that is rotating about the axis of symmetry of both, we have the following:

$$I_{total} = I_{\substack{solid\\sphere}} + I_{\substack{hollow\\cylinder}} = \tfrac{2}{5}M_s R_s^2 + M_c R_c^2$$

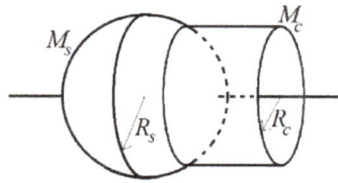

When using this deconstruction of a single complex object into simpler ones, we must remember that this only works when both objects share the same axis of rotation and rotate together as one object.

## Moments of Inertia of Off-Center Objects: The Parallel Axis Theorem

There will be problems to solve in which the rotating object is rotating about an axis that is parallel to but shifted to the side, away from the axis of symmetry. Examples of this would be a sphere rotating about an axis that is tangent to its surface, as illustrated below:

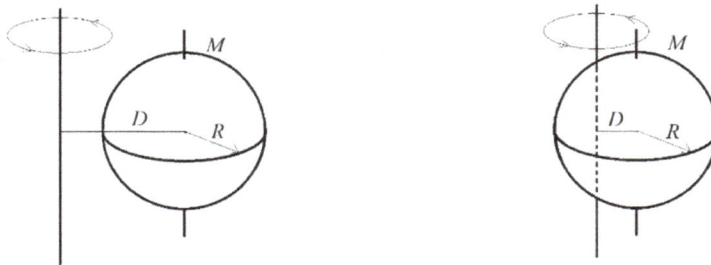

On the left, the sphere is rotating about an axis outside of its surface. On the right, the axis of rotation is through the body. In each case, the distance from the axis of symmetry to the rotation axis is labeled as $D$.

In these situations, the moment of inertia of the body undergoing off-axis rotation can be found by using what is called the Parallel Axis Theorem. Mathematically, the Parallel Axis Theorem is given by the following relationship:

$$I = I_{CM} + MD^2$$

Here, $I$ is the moment of inertia of the body about the non-symmetry axis, $I_{CM}$ is the moment of inertia of the object for rotation about its symmetry axis, $M$ is its mass, and $D$ is the distance from the symmetry axis to the axis about which it is rotating. As one would expect for a mass that is rotating off-axis, the further it is from the axis of rotation, the larger the moment of inertia. This is captured in the expression above in the term $MD^2$.

The Parallel Axis Theorem can also be useful in problems that include rolling motion. For these, the instantaneous pivot point is at the point of contact, with the length $D$ being between the pivot point and the symmetry axis of the object.

## Translational and Rotational Quantities

Generally, there is a one-to-one correspondence between the translational and rotational versions of the major quantities of mechanics and how they are used. These are summarized in table 6.4, for which the following mapping has been made:

TABLE 6.4  Translational and Rotational Versions of Quantities of Mechanics

| Quantity | Variable |
|---|---|
| inertia | $m \to I$ |
| acceleration | $a \to \alpha$ |
| velocity, speed | $v \to \omega$ |
| displacement, position | $x \to \theta$ |
| momentum | $p \to L$ |
| force | $F \to \Gamma$ |

Table 6.5 lists the full expressions for the various quantities of mechanics and shows the similarities between the translational and rotational forms. For the rotational version of the translational quantities that are scalar, such as work and energy, the similarities in the structure of the expressions are easy to see. For the vector quantities of torque and angular momentum, however, it is necessary to remember how to apply a cross-product to use them correctly.

TABLE 6.5  Mechanical Relations: Their Translational and Rotational Versions

| Quantity or Relation | Translational Version | Rotational Version |
|---|---|---|
| Momentum | $\vec{p} = m\vec{v}$ | $\vec{L} = I\vec{\omega} = \vec{r} \times \vec{p}$ |
| Newton's Second Law | $\sum \vec{F} = d\vec{p}/dt = m\vec{a}$ | $\sum_i \vec{\Gamma}_i = d\vec{L}/dt = I\vec{\alpha} = \sum_i (\vec{r}_i \times \vec{F}_i)$ |
| Kinetic Energy | $K = \frac{1}{2}mv^2 = p^2/2m$ | $K = \frac{1}{2}I\omega^2 = L^2/2I$ |
| Work | $dW = \vec{F} \cdot d\vec{r},\, W = \int \vec{F} \cdot d\vec{r}$ | $dW = \vec{\Gamma} \cdot d\vec{\theta},\, W = \int \vec{\Gamma} \cdot d\vec{\theta}$ |

### Application of Definitions of Rotational Forms of Mechanics Quantities

As is generally true, a picture can be beneficial to understanding how a concept should be applied. The following discussion is a reminder of how the rotational quantities of mechanics are defined and applied in problem-solving.

Torque: When applying Newton's Second Law in a rotational problem, the magnitude of the torque depends on how the force is applied in a way that it does not for translational problems. The illustration shows a force applied to a mass that is constrained to move in a circle about a point $P$. In the picture, notice that it is only the component of the force that is perpendicular to the line between where the force is applied to the pivot point that can cause a change in rotational speed. This is labeled as $\vec{F}_\perp$ for being perpendicular to the position vector from the pivot point to where the force is applied. In the case illustrated, the direction of the angular acceleration is such that it is out of the page when viewed from above as shown on the right:

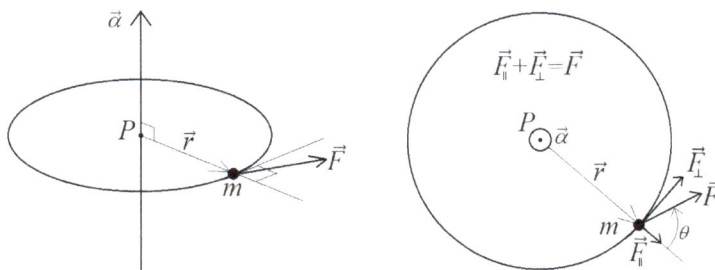

The relationship between torque, force, and where the force is applied is captured by the use of the cross-product, which provides the direction of the resultant torque vector and that only the component $\vec{F}_\perp$ contributes to the acceleration. In other words, $\vec{\Gamma} = \vec{r} \times \vec{F}$ and $|\vec{\Gamma}| = \Gamma = rF \sin\theta$.

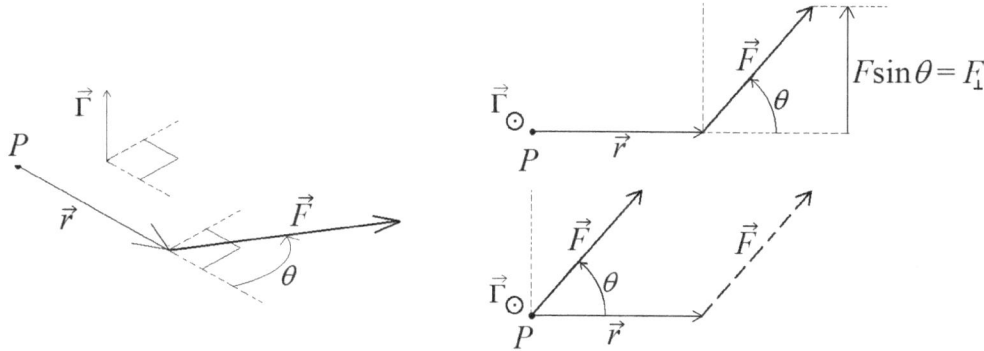

A common pitfall while solving torque problems is forgetting the vector nature of the cross-product and that it is an order-dependent operation. In other words, $\vec{C} = \vec{A} \times \vec{B}$, $|\vec{C}| = |\vec{A}||\vec{B}| \sin\theta_{\vec{A} \to \vec{B}}$, $\vec{C}$ is perpendicular to the plane defined by $\vec{A}$ and $\vec{B}$, and $\vec{A} \times \vec{B} = -(\vec{B} \times \vec{A})$.

Conceptually, one can think of the cross-product of two vectors as being a measure of how perpendicular they are to each other. To apply as much torque as possible with a force, we would want the force to be as perpendicular as possible to the line between the pivot point and the object. This amount of perpendicularity is captured by the sine function.[3]

Angular momentum: Similarly, the angular momentum is defined for a single particle as $\vec{L} = \vec{r} \times \vec{p}$, for which the same caveats about its vector nature and order of operations apply as for the torque.

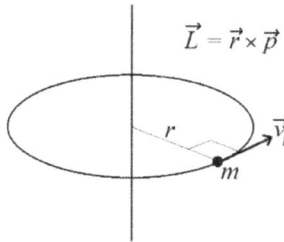

For the special case above of a single particle moving in a circle, the velocity vector, and hence the instantaneous translational momentum, is tangential to the circular path so that the angle between the position vector and the momentum vector is 90° with the result from the cross-product definition we get:

$$L = |\vec{r} \times \vec{p}| = rp \sin 90^0 = rp(1) = rmv = mvr$$

In terms of $\omega$ and $I$, the magnitude of the angular momentum can be rewritten as

$$L = mvr = m(r\omega)r = (mr^2)\omega = I\omega.$$

Angular momentum of a body in straight-line motion: With regard to solving problems that include straight-line motion and rotational motion, such as when two objects collide where one is moving along a straight path, the translational motion of the object moving in a straight line can also be described using angular momentum.

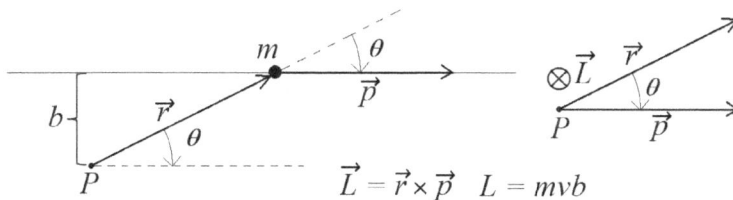

---

3   In contrast, the dot-product, $D = \vec{A} \cdot \vec{B} = AB \cos\theta_{\vec{A} \to \vec{B}}$, is a measure of how parallel two vectors are to each other, which is captured by the cosine function.

For the object of mass $m$ passing within a distance $b$ of a point $P$ at a constant velocity, the cross-product, $\vec{L} = \vec{r} \times \vec{p}$, indicates that the angular momentum vector is into the page with magnitude $L = mvr \sin\theta$. Because the product of the sine of the angle and the distance to the point of reference is a constant, $b = r \sin\theta$, the magnitude of the angular momentum is $mvb$ and thus constant, as is the translational momentum of the particle with magnitude $mv$.

<u>Work done by a torque</u>: As with translational motion, there is also a rotational version of the Work-Kinetic Energy Theorem. The derivation is done following the same steps as for the translational form, by starting with Newton's Second Law and integrating the effect of the torque for a given rotational displacement. This is normally done nicely in standard textbooks so that only the result is given below where the rotational kinetic energy is given by $K_R = \frac{1}{2} I\omega^2$.

$$\Delta K = \sum W_R = \tfrac{1}{2}I\omega_f^2 - \tfrac{1}{2}I\omega_i^2 \quad \text{where} \quad W_R = \int_i^f \vec{\Gamma} \cdot d\vec{\theta}$$

If the direction of the force applied to a rotating body is perpendicular to the axis of rotation, then $\vec{\Gamma}$ and $d\vec{\theta}$ both point along the axis of rotation, and therefore being parallel, $\vec{\Gamma} \cdot d\vec{\theta} = \Gamma d\theta$ so that $W_R = \int_i^f \vec{\Gamma} \cdot d\vec{\theta} = \int_i^f \Gamma d\theta$, which has the same form as the translational version $W_T = \int_i^f \vec{F} \cdot d\vec{r} = \int_i^f F dr$.

We can deduce the units for rotational work, and rotational Kinetic energy is Joules, as it is for the translational versions:

$$\left[\sum W_R\right] = \left[\tfrac{1}{2}I\omega_f^2 - \tfrac{1}{2}I\omega_i^2\right] = \left[\tfrac{1}{2}I\omega^2\right] = \left[\tfrac{1}{2}\right][I][\omega^2] = 1(\text{kg}\cdot\text{m}^2)(1/\text{s})^2 = \text{kg}\cdot\text{m}^2/\text{s}^2 = \text{Joules}$$

## Summary of Rotational Units

With these new concepts, it is worthwhile to summarize the units of the rotational terms.

TABLE 6.6  Summary of Rotational Units

| Parameter | Parameter or Equation | Units |
|---|---|---|
| Angular Displacement | $\Delta\theta$ or $\theta$ | 1 (unitless) |
| Angular Velocity | $\vec{\omega} = d\vec{\theta}/dt$ | 1/s |
| Angular Acceleration | $\vec{\alpha} = d\vec{\omega}/dt$ | 1/s$^2$ |
| Moment of Inertia | $I = \beta MR^2$ | kg$\cdot$m$^2$ |
| Torque | $\vec{\Gamma} = \vec{r} \times \vec{F} = I\vec{\alpha}$ | $(\text{m})\cdot(\text{kg}\cdot\text{m}/\text{s}^2) = \text{kg}\cdot\text{m}^2/\text{s}^2$ $= \text{N}\cdot\text{m}$ |
| Angular Momentum | $\vec{L} = \vec{r} \times \vec{p} = I\vec{\omega}$ | $(\text{m})\cdot(\text{kg}\cdot\text{m}/\text{s}) = \text{kg}\cdot\text{m}^2/\text{s}$ |
| Work and Kinetic Energy | $\sum W_R = \tfrac{1}{2}I\omega_f^2 - \tfrac{1}{2}I\omega_i^2 = \int_i^f \left(\sum \vec{\Gamma} \cdot d\vec{\theta}\right)$ | $(\text{kg}\cdot\text{m}^2)(1/\text{s}^2) = \text{kg}\cdot\text{m}^2/\text{s}^2$ $= \text{N}\cdot\text{m} = \text{J}$ |

A curious artifact of the units used for rotational dynamics is that <u>the units for torque happen to be the same as those for work and energy</u>: N·m = J. Another accident of units is that <u>the units for angular velocity are the same as for frequency</u>, 1/s. As you do your physics problems, keep context in mind and remember that even though something has the units you expect, it is still important to check that they originate in the way you expect.

## The Problem-Solving Steps for Rotational Dynamics Problems

Solving problems using the rotational versions of Newton's Second Law, the Work-Kinetic Energy Theorem, Conservation of Energy, or Conservation of Momentum is fundamentally the same as for when solving translational dynamics problems.

## For Each Type of Problem, Whether Translational or Rotational, the Basic Steps are the Same

- *Abstract the problem* into a more mathematical form
- *Identify the physics* principle and its mathematical form to be used
- *Apply the mathematical form of the principle* to the problem to be solved
- *Use mathematics*, algebra and/or calculus to simplify the starting mathematical expression into the desired form
- *Check the final expression* with unit analysis and limiting cases
- *Evaluate the final expression* numerically and check the physicalness of the answer

For each of the key physics principles, no specialized problem-solving steps are needed because, as stated above, the steps are basically the same irrespective if the problem is translational or rotational.

## Newton's Second Law with Rotational Motional

As was done for using Newton's Second Law for translational problems, the rotational version of Newton's Second Law is applied similarly so that the instructions for its application are generalized as follows:

- For each object with inertia being *acted upon by a force*, sum each of the forces vectorially and equate the sum to the product of the mass and the translational acceleration: $\sum \vec{F} = m\vec{a}$
- For each object with inertia being *acted upon by a torque*, sum each of the torques vectorially and equate the sum to the product of the moment of inertia and the angular acceleration: $\sum \vec{\Gamma} = I\vec{\alpha}$

## Work-Kinetic Energy Theorem with Rotational Motion

Similar to the more generalized rules for applying Newton's Second Law, the Work-Kinetic Energy Theorem is generalized as follows:

- $\Delta K = \sum W$ where
  - for a *translational motion*:

$$\Delta K = \tfrac{1}{2}mv_f^2 - \tfrac{1}{2}mv_i^2 \text{ and } \sum W = \left(\sum \vec{F}\right)\cdot\Delta\vec{r} \text{ or } \sum W = \int \left(\sum \vec{F}\right)\cdot d\vec{r}$$

  - for a *rotational motion*:

$$\Delta K = \tfrac{1}{2}I\omega_f^2 - \tfrac{1}{2}I\omega_i^2 \text{ and } \sum W = \left(\sum \vec{\Gamma}\right)\cdot\Delta\vec{\theta} \text{ or } \sum W = \int \left(\sum \vec{\Gamma}\right)\cdot d\vec{\theta}.$$

## Conservation of Energy with Rotational Motion

For conservation of energy, it is in the kinetic energy terms where both any translational and rotational kinetic energies are included. Otherwise, the setup and starting equation are the same.

- $E_f = E_i + \sum W_{nc}$ where $E = \sum K + \sum U = \left(\sum K_{Trans} + \sum K_{Rot}\right) + \sum U$.

## Conservation of Momentum; Translational and Rotational

Generally, the expressions for the law of Conservation of Momentum and law of Conservation of Angular Momentum are not combined into one single expression as was done for Conservation of Energy. As such, each is applied separately to describe the translational and rotational aspects of the motion and interactions. The results are then combined mathematically as necessary. However, what is the same is the basic steps to the application of each of the two momentum laws.

- The sum of the momenta in the initial state is equal to the sum of the momenta in the final state:

  - Translation: $\left(\sum_i \vec{p}_i\right)_{initial} = \left(\sum_j \vec{p}_j\right)_{final}$

  - Rotation: $\left(\sum_i \vec{L}_i\right)_{initial} = \left(\sum_j \vec{L}_j\right)_{final}$

# Newton's Second Law: Problem-Solving Steps (Translational or Rotational)

<u>Hallmarks</u>: Forces or torques applied to one or more bodies and the acceleration (or lack of acceleration) of those bodies mentioned.

**Newton's Second Law**

$$\sum \vec{F} = m\vec{a} \quad \text{or} \quad \sum \vec{\Gamma} = I\vec{\alpha}$$

## Steps

1. **Identify components of system\***

   - *Abstract the word problem* into mathematical expressions with a figure

   - *Apply the physics principle* by drawing a free-body diagram for each object upon which forces or torques act (to aid in writing out full vector equation starting with $\sum \vec{F} = m\vec{a}$ and $\sum \vec{\Gamma} = I\vec{\alpha}$)\*

   - *Define a coordinate system* for each object that makes defining the vectors simple

2. **Adapt and apply Newton's Second Law to the problem\***

   - *Write out the vector form of the equation* for Newton's Second Law for each object, using vector variables

   - *Rewrite vector equation in component form* using unit vectors for each of the vectors in the $\sum \vec{F} = m\vec{a}$ and/or $\sum \vec{\Gamma} = I\vec{\alpha}$ equation for each object

   - *Collect the terms* of vector equation(s) by unit vector

   - *Separate out the scalar equations* for each axis for each body's $\sum \vec{F} = m\vec{a}$ and/or $\sum \vec{\Gamma} = I\vec{\alpha}$ equation

3. **Identify any other relationships needed**

   - *Check* if you have the same number of equations as unknown variables

   - If necessary, *identify more data and/or equations* so that the number of equations equals the number of unknowns and the system can be solved

   - *Write out the final set* of equations

4. **Do algebra to solve equations for parameter of interest**

   - *Use only variables*, not numerical values

5. **Check "physicalness" of final relationship**

   - *Perform unit analysis*

   - *Check predictions* of limiting behavior

6. **Evaluate expression using data** (if provided)

   - *Check physicalness and reasonableness* of the answer

*\*These are the key steps for this type of problem.*

# Work-Kinetic Energy: Problem-Solving Steps (Translational and/or Rotational)

<u>Hallmarks</u>: There are identifiable initial and final speeds for the object on which forces or torques are acting, and the forces or torques described are acting over a distance along a path or some displacement. Time is not explicitly noted:

---

**Work-Kinetic Energy Theory**

$$\sum W = \Delta K = K_f - K_i$$

where $\sum W = \int_{\vec{r}_i}^{\vec{r}_f} \left(\sum \vec{F}\right) \cdot d\vec{r}$ or $\sum W = \int_{\vec{\theta}_i}^{\vec{\theta}_f} \left(\sum \vec{\Gamma}\right) \cdot d\vec{\theta}$

$K_f = \frac{1}{2}mv_f^2$ and $K_i = \frac{1}{2}mv_i^2$ and/or $K_f = \frac{1}{2}I\omega_f^2$ and $K_i = \frac{1}{2}I\omega_i^2$

---

## Steps

1. **Identify components of system\*** (A sketch is very useful for this)

   - *Draw a diagram* for the initial and final states along with a free-body diagram and the displacement vector (translation or rotation) to aid in the initial setup of the problem

   - *Define the key variables* used with analytic expressions, including zeros for any coordinate systems

2. **Adapt the general principle to the specific situation to be described\***

   - *For the Work term*, identify if the net force or net torque is constant or if it acts at a fixed angle to the path of the object

     - If force/torque is not constant or if angle of force to path is not constant, use integral form for work

     - If force/torque is constant and if angle between force and path is constant, use $W = \vec{F} \cdot \vec{d}$ or $W = \vec{\Gamma} \cdot \Delta\vec{\theta}$

   - *For Kinetic Energy terms*, fill in expression for kinetic energy, $\frac{1}{2}mv^2$ or $\frac{1}{2}I\omega^2$, for initial and final states. Use zero for either that corresponds to the object being at rest

3. **Identify and apply any other important relationships**

   - *Use Newton's Second Law* as necessary to find relationships between magnitudes of forces and torques (e.g., Find an expression for the normal force so that the friction force, $f = \mu n$, can be expressed properly)

   - *Fill in any special values or relationships* for variables; e.g., $y_f = h$, $v_i = 2v_f$, or simplification s of the dot-product

4. **Do algebra and/or integration to solve equations for parameter of interest**

   - *Use only variables*, not numerical values

5. **Check "physicalness" of final symbolic relationship**

   - *Perform unit analysis*

   - *Check predictions* of limiting behavior

6. **Evaluate expression using data** (if provided)

   - *Check physicalness and reasonableness* of answer

*\*These are the key steps for this type of problem.*

# Conservation of Energy: Problem-Solving Steps (Translational and/or Rotational)

<u>Hallmarks</u>: Work (forces and distances), initial and final speeds, initial and final positions are mentioned.

**Law of Conservation of Energy for Mechanics**

$$E_f = E_i + \sum W_{nc} \quad \text{or} \quad \left(\sum K_f + \sum U_f\right) = \left(\sum K_i + \sum U_i\right) + \sum W_{nc}$$

## Steps

1. **Identify components of system\*** (A sketch is very useful for this)
   - *Count number of objects moving* transitionally and rotationally (for changes in kinetic energy)
   - *Count number of objects changing position* (for changes in potential energy)
   - *Count number of non-conservative interactions* of objects with other objects (for work done on each object)
   - *Define the key variables* used with analytic expressions, including zeros for any coordinate systems

2. **Fill in placeholders for each energy term\***
   - For each moving object, *add one initial and one final Kinetic Energy* term on *each* side of the equation using $K_T$ or $K_R$ as appropriate
   - For each object changing position, *add one initial and one final Potential Energy term* on *each* side of the equation using $U$ as appropriate
   - For each non-conservative work term, *add one Work term* on *initial* side of the equation

3. **For each energy term, substitute the appropriate expression**
   - *Kinetic energy*:
     - If initial or final speed is zero, substitute zero
     - If initial or final speed is not zero, substitute $\frac{1}{2}mv^2$ or $\frac{1}{2}I\omega^2$
   - *Potential energy*:
     - If an initial or final potential energy is zero, substitute zero
     - If an initial or final position is such that the potential energy is non-zero, substitute in full expression, for example, $mgy$ or $\frac{1}{2}kx^2$
   - *Non-conservative work*:
     - For constant forces or torques, $W = \vec{F} \cdot \vec{d} = Fd\cos\theta_{\vec{F},\vec{d}}$ or $W = \vec{\Gamma} \cdot \Delta\vec{\theta}$
     - For non-constant forces or torques, $W = \int \vec{F} \cdot d\vec{r}$ or $W = \int \vec{\Gamma} \cdot d\vec{\theta}$
     - If needed, use Newton's Second Law to find expression for the force $\vec{F}$ or torque $\vec{\Gamma}$
   - *Fill in any special values* for variables; e.g., $y_f = h$, $v_i = v_f$

4. **Do algebra to solve equations for parameter of interest**
   - *Use only variables*, not numerical values

5. **Check "physicalness" of final relationship**
   - *Perform unit analysis*
   - *Check predictions* of limiting behavior

6. **Evaluate expression using data** (if provided)
   - *Check physicalness and reasonableness* of answer

*\*These are the key steps for this type of problem.*

# Conservation of Momentum: Problem-Solving Steps (Translational and/or Rotational)

<u>Hallmarks</u>: Identifiable initial and final states for a closed system in an inertial reference frame that can be defined in terms of momenta or velocities and masses or moments of inertia of the objects. The nature of the forces or torques might or might not be apparent.

---

**Laws of Conservation of Translational and Rotational Momentum**

$$\left(\sum_i \vec{p}_i\right)_{init} = \left(\sum_j \vec{p}_j\right)_{final} \quad \text{or} \quad \left(\sum_i \vec{L}_i\right)_{init} = \left(\sum_j \vec{L}_j\right)_{final}$$

---

## Steps

1. **Identify the components of the system\***
   - *Abstract the problem* into figures and mathematical form
     - *Draw a figure* that describes the initial and final states, including vectors for momentum or velocities and identify the coordinate system to be used
     - *Define the key variables* or relationship between parameters used for the system components
     - *Identify any other relationships* that will or could be used based on the physics, e.g., the Law of Conservation of Energy.

2. **Apply the concept of the Law of Conservation of Momentum\***
   - *Write down general form for the Law* of Conservation of Momentum appropriate for the problem
   - *For initial state side of equation, add vectorially the momentum* of each object
   - *For final state side of equation, add vectorially the momentum* of each object

3. **Rewrite the vector equation as a set of scalar equations**
   - *Rewrite vectors in component form and collect terms by unit vector.* For 1-dimensional/rotational problems, use + and – signs to designate direction
   - *Identify and write out the scalar equations* associated with each unit vector

4. **Do algebra to solve for the parameter of interest**
   - *Check if number of equations equals number of unknowns*
     - *If there are equal numbers of each, proceed*
     - *If not, identify any additional relationships* that might be necessary that were not included in earlier steps
   - *Solve for variable/parameter of interest*

5. **Check "physicalness" of final symbolic relationship**
   - *Perform unit analysis*
   - *Check predictions and special cases* for physicalness

6. **Evaluate expression using data** (if provided)
   - *Check physicalness and reasonableness* of answer

*\*These are the key steps for this type of problem.*

# EXAMPLE 6.1

A weight of mass $m_1$ hangs from a string that is wrapped around a spool of radius $b$ on a frictionless axle that has a mass of $m_2$ and moment of inertia about its axle of $I = \beta m_2 b^2$. Find an expression for the rotational acceleration and the speed of the spool after the weight has dropped a distance $h$ from rest in terms of $\beta$, $m_1$, $m_2$, $b$, $h$.

## SOLUTION:

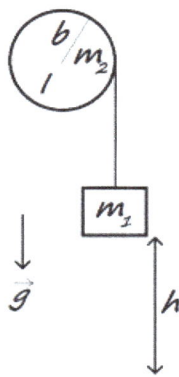

$\alpha = ?,$

$\omega_f = ?$ after falling a distance $h$

spool

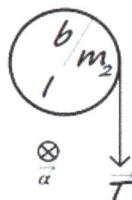

## ANNOTATION/COMMENT (STEP #):

Forces, torques, accelerations, masses, and moments of inertia: Use of Newton's Second Law is appropriate.

**Draw diagram to help identify all important parts. (1)**

**Identify the raw data of the problem and the key request being made. (1)**

**Draw a free body diagram and acceleration vector for each object on which forces or torques act (1)**

Note: The coordinate system used for each object does not need to be the same.

| SOLUTION: | ANNOTATION/COMMENT (STEP #): |
|---|---|

<u>weight</u>

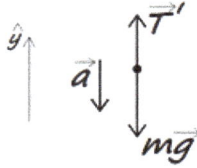

**Draw the free-body diagram to identify the forces acting on the first of the two bodies and include the expected acceleration vector (1)**

<u>spool</u>

$$\sum \vec{\Gamma} = I\vec{\alpha}$$

$$\boxed{-Tb = I\alpha}$$

**Translate Newton's Second Law, $\sum \vec{F} = m\vec{a}$ or $\sum \vec{\Gamma} = I\vec{\alpha}$, for each object into scalar form (2)**

<u>Scalar equation for the spool</u>

<u>weight</u>

$$\sum \vec{F} = m\vec{a}$$

$$\vec{T}' + m_1\vec{g} = m_1\vec{a}$$

$$\boxed{T' - m_1 g = -m_1 a}$$

Note: It is this step that negative signs that denote direction are normally included.

<u>Scalar equation for the weight</u>

$$a = b\alpha$$

$$\left| \vec{T}' \right| = \left| \vec{T} \right| \Rightarrow T' = T$$

$$I = \beta mb^2$$

**Identifying other relationships that appear to be useful for describing the system (3)**

$$T - m_1 g = -m_1(b\alpha)$$
$$T - m_1 g = -m_1 b\alpha$$

Doing the preliminary substitutions:

$T' = T$ and $a = b\alpha$ into equation for the weight

$$-Tb = \beta m_2 b^2(-\alpha)$$
$$Tb = \beta m_2 b^2 \alpha$$

$I = \beta m_2 b^2$ into equation for the spool

$$\boxed{\begin{array}{l} T - m_1 g = -m_1 b\alpha \\ \qquad Tb = \beta m_2 b^2 \alpha \end{array}}$$

<u>Final set of equations from Newton's Second Law equations</u>

unknowns: $\alpha$, $T$

## SOLUTION:

$$\boxed{\begin{aligned} T - m_1 g &= -m_1 b\alpha \\ Tb &= \beta m_2 b^2 \alpha \end{aligned}}$$

unknowns: $\alpha$, $T$

$$Tb = \beta m_2 b^2 \alpha$$
$$T = \beta m_2 b\alpha$$

$$(\beta m_2 b\alpha) - m_1 g = -m_1 b\alpha$$
$$\beta m_2 b\alpha + m_1 b\alpha = m_1 g$$
$$(\beta m_2 b + m_1 b)\alpha = m_1 g$$
$$\alpha = \frac{m_1 g}{\beta m_2 b + m_1 b}$$

$$\boxed{\alpha = \frac{g}{b(\beta(m_2 / m_1) + 1)}}$$

$$[\alpha] = \left[\frac{g}{b(\beta(m_2 / m_1) + 1)}\right]$$
$$= \frac{[g]}{[b]([\beta]([m_2 / m_1]) + [1])}$$
$$= \frac{(m/s^2)}{(m)(1)(1(\cancel{kg}/\cancel{kg}) + 1)}$$

$$[\alpha] = 1/s^2 \checkmark$$

## ANNOTATION/COMMENT (STEP #):

**Check if number of scalar equations equals number of unknowns (3)**

Note: Checking if the number of equations equals the number of unknowns can often be done by inspection for simpler systems.

Although not provided, variables by which the final expressions to be defined are counted as knowns.

**Perform algebra to find angular acceleration (4)**

Final expression for acceleration

**Perform unit analysis (5)**

Note: This can be an easy step to do by inspection without writing out. It should always be done as a quick check to see if any errors have occurred.

Units are $1/s^2$ as expected

| SOLUTION: | ANNOTATION/COMMENT (STEP #): |
|---|---|

**Check physicalness (5)**

For $m_2 \ll m_1$ or $(m_2/m_2) \ll 1$

$$\alpha = \frac{g}{b(\beta(m_2/m_1)+1)}$$

$$\sim \frac{g}{b(\beta(0)+1)}$$

$$\sim g/b$$

$$b\alpha \sim a \sim g \checkmark$$

Checking expectations for if the mass of the spool is negligibly small.

The weight would free-fall at about $1\,g$, as would be expected if the spool had little to no inertia.

For $m_2 \gg m_1$ or $(m_1/m_2) \ll 1$

$$\alpha = \frac{g}{b(\beta(m_2/m_1)+1)}$$

$$\sim \frac{g}{b(\beta(m_2/m_1))}$$

$$\sim \left(\frac{m_1}{m_2}\right)\frac{g}{b\beta}$$

$$\alpha \sim 0 \checkmark$$

Checking expectations for the case that the weight has little to no mass.

The spool will not accelerate because the torque on it is about zero, as expected.

$$\boxed{\alpha = \frac{g}{b(\beta(m_2/m_1)+1)}}$$

Based on unit analysis and limiting case analysis, the final expression is very likely correct.

**152** Physics Problem-Solving Techniques for Understanding and Success in First Year Mechanics

| **SOLUTION:** | **ANNOTATION/COMMENT (STEP #):** |
|---|---|
| | Find the expression for final angular speed |
| $\omega_f = ?$ | **Combine acceleration and angular displacement to find final angular velocity (3)** |
| $\omega_i = 0$, $\theta_i = 0$ and $\theta_f = h/b$ | |
| | Kinematics relationship identified to use. Could use others but would require also finding expression for the time it takes for the weight to fall. |
| $\boxed{2\alpha(\theta_f - \theta_i) = \left(\omega_f^2 - \omega_i^2\right)}$ | |
| $2\alpha(\theta_f - \theta_i) = (\omega_f^2 - \omega_i^2)$ | **Doing algebra (4)** |
| $2\alpha(\theta_f - 0) = (\omega_f^2 - 0)$ | |
| $2\alpha\theta_f = \omega_f^2$ | |
| $\omega_f^2 = 2\alpha\theta_f$ | |
| $= 2\left(\dfrac{m_1 g}{b(\beta m_2 + m_1)}\right)\left(\dfrac{h}{b}\right)$ | Substituting expression for the angular acceleration $\alpha$. |
| $\omega_f^2 = \dfrac{2m_1 gh}{b^2(\beta m_2 + m_1)}$ | |
| $\boxed{\omega_f = \dfrac{1}{b}\left(\dfrac{2m_1 gh}{(\beta m_2 + m_1)}\right)^{1/2}}$ | <u>Expression for the final speed</u> |

# EXAMPLE 6.2

A solid ball is rolling down a ramp inclined at an angle of 25°. Its mass is 450 g and its radius is 6.5 cm. What is its angular acceleration as it rolls? What is its translational acceleration down the ramp? If it rolls a distance of 20 cm, what is its final speed if it was released from rest?

**SOLUTION:**

**ANNOTATION/COMMENT (STEP #):**

Forces, torques, accelerations, masses, and moments of inertia: Use of Newton's Second Law is appropriate.

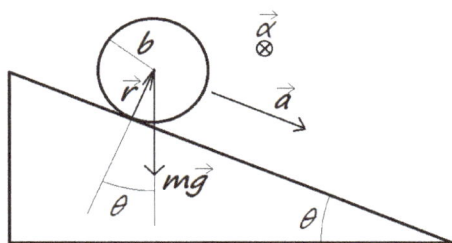

**Sketch of system to help in identifying the key parts (1)**

$m = 0.45$ kg, $b = 6.5$ cm,
$\theta = 25°$, shape: solid sphere,
$\omega_i = 0$, $d = 20$ cm
$\alpha = ?$, $\omega_f = ?$

**Data for parts of system (1)**

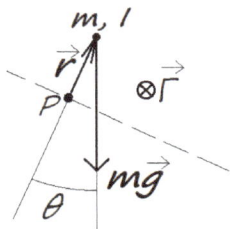

**Free-body diagram for applying Newton's Second Law using torque. (1)**

Instantaneous axis of rotation is at point of contact, $P$, of ball with ramp:

$$\sum \vec{\Gamma} = I\vec{\alpha}$$
$$\vec{r} \times m\vec{g} = I\vec{\alpha}$$
$$-bmg \sin\theta = -I\alpha$$
$$bmg \sin\theta = I\alpha$$

**Applying Newton's Second Law (2)**

Torque is negative because it points into the page as is the angular acceleration

**Additional relationship needed (3)**

$$I = I_{CM} + MD^2 = I_{sph} + mb^2$$
$$= \frac{2}{5}mb^2 + mb^2 = \frac{7}{5}mb^2$$

Applying parallel axis theorem for rotation of ball about point $P$.

153

| SOLUTION: | ANNOTATION/COMMENT (STEP #): |
|---|---|

**Doing algebra (4)**

$$bmg \sin\theta = I\alpha$$

$$\alpha = \frac{bmg \sin\theta}{I}$$

$$= \frac{bmg \sin\theta}{\left(\frac{7}{5}mb^2\right)}$$

Substituting expression for moment of inertia

$$= \left(\frac{5}{7}\right)\left(\frac{b\,mg \sin\theta}{mb^2}\right)$$

$$\boxed{\alpha = \left(\frac{5}{7}\right)\frac{g \sin\theta}{b}}$$

Final expression for angular acceleration

**Checking units (5)**

$$[\alpha] = \left[\left(\frac{5}{7}\right)\frac{g \sin\theta}{b}\right]$$

Note: Doing unit analysis is an easy way to check if any errors were made before moving on. Only for complicated expressions is writing it out ever necessary.

$$= \frac{5}{7}\frac{[g]}{[b]}[\sin\theta]$$

$$= (1)\left(\frac{m/s^2}{m}\right)(1) = \left(\frac{\cancel{m}/s^2}{\cancel{m}}\right)$$

$$[\alpha] = 1/s^2 \checkmark$$

Units are $1/s^2$ as expected for angular acceleration.

**Checking Expectations (5)**

For $\theta \rightarrow 90°$

$$\alpha = \left(\frac{5}{7}\right)\frac{g \sin\theta}{b} \rightarrow \left(\frac{5}{7}\right)\frac{g}{b} \checkmark$$

For $\theta \rightarrow 90°$, the magnitude of the angular acceleration is maximized as expected.

For $\theta \rightarrow 0°$

$$\alpha = \left(\frac{5}{7}\right)\frac{g \sin\theta}{b} \rightarrow 0 \checkmark$$

For $\theta \rightarrow 0°$, acceleration goes to zero as expected.

Note: Checking limiting behavior is a good way to check that the problem has been correctly or not.

**SOLUTION:**

$$\alpha = \left(\frac{5}{7}\right)\frac{g\sin\theta}{b}$$

$$= \left(\frac{5}{7}\right)\left(9.80\,\frac{m}{s^2}\right)\left(\frac{\sin 25^\circ}{0.065\,m}\right)$$

$$= 45.5\,\frac{1}{s^2}$$

$$\boxed{\alpha = 46\,\frac{rad}{s^2}}$$

$$a = b\alpha$$

$$= (6.5\,\cancel{cm})\left(-46\,\frac{rad}{s^2}\right)\left(\frac{m}{100\,\cancel{cm}}\right)$$

$$\boxed{a = 3.0\,\frac{m}{s^2}}$$

$$\underline{v_f = ?}$$

$$2a(x_f - x_i) = \left(v_f^2 - v_i^2\right)$$

$$(v_f^2 - 0) = 2a(x_f - 0)$$

$$v_f^2 = 2ax_f$$

$$v_f = \pm\sqrt{2ax_f}$$

$$= \pm\sqrt{(2)\left(3.0\,\frac{m}{s^2}\right)(0.20\,m)}$$

$$v_f = \pm 1.1\,\frac{m}{s}$$

$$\boxed{v_f = 1.1\,\frac{m}{s}}$$

**ANNOTATION/COMMENT (STEP #):**

**Calculating the numerical value of answer (6)**

Final expression for the angular acceleration.

Sign is correct for rolling downhill and the magnitude of 7.3 rev/s² is reasonable.

**Calculating the translational acceleration (6)**

Note: Radians is unitless, so it is dropped in the conversion to translational units.

Final value for the magnitude of the translational acceleration

Value is a fraction of g with the correct sign for down the ramp.

**Using kinematics to find final translational speed (3-6)**

Note: When using an expression for which a value is squared, the answer can have the incorrect sign.

Final mathematical answer

The only physical answer is the magnitude of the velocity.

Final physical answer

# EXAMPLE 6.3

A wheel of mass 10 kg that can be modeled as having all its mass between radii of 65 cm and 75 cm has a narrow brake pad applied to its
outer edge. The friction coefficient between the brake pad and the wheel
is 0.85. The wheel initially rotates at a speed of +30 rpm and comes to
rest after 30 revolutions when the brake is applied. What was the force
applied to the brake?

**SOLUTION:**

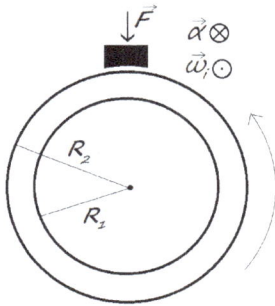

$w_i = +30$ rpm, $w_f = 0$,

$\Delta\theta = +30$ rev., $\mu = 0.85$,

$R_1 = 0.65$ m, $R_2 = 0.75$ m,

$M = 10$ kg  $F = ?$

$$\Delta K = \sum W$$

$$K_f - K_i = W_f$$

$$0 - K_i = \int_i^f \vec{\Gamma} \cdot d\vec{\theta}$$

$$-K_i = \vec{\Gamma} \cdot \Delta\vec{\theta}$$

$$-K_i = \Gamma \, \Delta\theta$$

$$-\tfrac{1}{2} I w_i^2 = \Gamma \, \Delta\theta$$

**ANNOTATION/COMMENT (STEP #):**

Change in speed and force applied over a displacement: Use of
Work-Kinetic Energy is appropriate.

**Diagram to identify the key parameters of the system (1)**

**The data for the problem (1)**

**Applying Work-Kinetic Energy Theorem for a rotational
system for which the force associated with the work is
constant (2)**

Because torque is constant, using non-integral version of work.
Because torque vector is on same axis as displacement, i.e., they
are parallel, we can drop the vectors and dot-product.

| SOLUTION: | ANNOTATION/COMMENT (STEP #): |
|---|---|
| | **Identifying other terms needed to solve for the force required to stop the wheel (3)** |
| $\tau = \left|\vec{\tau}\right| = \left|\vec{R}_2 \times \vec{f}\right| = -R_2 f$ | Expression for the magnitude of the torque with correct sign (into page) |
| $f = \mu n$ | Expression for friction force |
| $I = \frac{1}{2} M \left( R_1^2 + R_2^2 \right)$ | Expression for moment of inertia of the wheel |
| | Using Newton's Second Law to identify an expression for the force $F$ |
| | Free-body diagram |
| $\sum \vec{F} = m\vec{a}$ $\vec{F} + \vec{n} + \vec{f} = m\vec{a}$ $-F\hat{j} + n\hat{j} + f\hat{i} = ma\,\hat{i}$ | Applying Newton's Second Law |
| $\rightarrow \begin{cases} x)\ f = ma \\ y)\ -F + n = 0 \end{cases}$ | Scalar equations from vector equation |
| $\Rightarrow \boxed{F = n}$ | Force applied has the same magnitude as the normal force. (Not surprising, but it is worth showing the steps.) |
| $-\frac{1}{2} I \omega_i^2 = \tau \Delta\theta$ $-\frac{1}{2} I \omega_i^2 = (-f R_2) \Delta\theta$ $-\frac{1}{2} \left[ \frac{M}{2} \left( R_1^2 + R_2^2 \right) \right] \omega_i^2 = -(\mu F) R_2 \Delta\theta$ $\left( \frac{M}{4} \right) \left( R_1^2 + R_2^2 \right) \omega_i^2 = \mu F R_2 \Delta\theta$ | **Doing algebra (4)** |
| $\boxed{F = \dfrac{M \left( R_1^2 + R_2^2 \right) \omega_i^2}{4 \mu R_2 \Delta\theta}}$ | Final expression for the force |

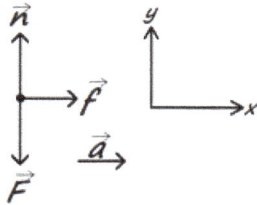

| SOLUTION: | ANNOTATION/COMMENT (STEP #): |
|---|---|

**Checking units (5)**

$$[F] = \left[\frac{M\left(R_1^2 + R_2^2\right)\omega_i^2}{4\mu R_2 \,\Delta\theta}\right]$$

$$= \frac{[M]\left[R_1^2 + R_2^2\right]\left[\omega_i^2\right]}{[4][\mu][R_2][\Delta\theta]}$$

$$= \frac{(kg)(m^2 + m^2)(1/s^2)}{(1)(1)(m)(1)}$$

$$= \frac{kg \cdot m^2}{\cancel{m} \cdot s^2} = kg\,\frac{m}{s^2}$$

$$[F] = N \; \checkmark$$

Note: This step should always be done, if even only by inspection. If the units are not correct here, that is a clue that there was an algebra mistake earlier that should be fixed before going on.

Units are Newtons, as expected.

**Checking predicted behavior (5)**

$$\underline{\Delta\theta \to \infty \text{ or } \Delta\theta \to 0:}$$

$$F = \frac{M\left(R_1^2 + R_2^2\right)\omega_i^2}{4\mu R_2 \Delta\theta} \sim \frac{1}{\Delta\theta} \; \checkmark$$

The amount of force needed to stop the wheel is inversely proportional to the displacement needed to stop it, as expected.

Note: Checking expectations should always be done even if not written out to check your work.

$$F = \frac{M\left(R_1^2 + R_2^2\right)\omega_i^2}{4\mu R_2 \Delta\theta} \sim \omega_i^2 \; \checkmark$$

The faster the initial speed, the greater the force required to stop it as expected.

$$\underline{M \to \infty \text{ or } M \to 0:}$$

$$F = \frac{M\left(R_1^2 + R_2^2\right)\omega_i^2}{4\mu R_2 \Delta\theta} \sim M \; \checkmark$$

The greater (smaller) the mass of the wheel, the greater (smaller) the force required to bring it to a stop.

$$\underline{R_2 \to \infty}$$

$$F = \frac{M\left(R_1^2 + R_2^2\right)\omega_i^2}{4\mu R_2 \Delta\theta} \sim C + DR_2 \; \checkmark$$

The greater the outside diameter, the greater the rotational inertia and the more difficult it is to bring the wheel to a stop.

| SOLUTION: | ANNOTATION/COMMENT (STEP #): |
|---|---|

**Evaluating final expression numerically (6)**

$$F = \frac{M\left(R_1^2 + R_2^2\right)\omega_i^2}{4\mu R_2 \,\Delta\theta}$$

$$= \frac{(10\ kg)(0.65\ m)^2 + (0.75\ m)^2)}{4(0.85)(0.75\ m)(30\ rev)}$$

$$\cdot \left(30\,\frac{rev}{min}\right)^2$$

$$= 116\,\frac{kg\cdot m^2\cdot rev^2}{rev\cdot m\cdot min^2}$$

Note: Sometimes it is fine to leave the units of revolution (rev) to see if they cancel. This can save some extra work.

$$= 116\,\frac{kg\cdot m^2\cdot rev^2}{rev\cdot m\cdot min^2}\left(\frac{1\,min}{60\ s}\right)^2\left(\frac{2\pi}{rev}\right)$$

$$= 116\,\frac{kg\cdot m^2\cdot rev^2}{rev\cdot m\cdot min^2}\,\frac{1\,min^2}{3600\ s^2}\left(\frac{2\pi}{rev}\right)$$

$$\boxed{F = 0.20\ N}$$

Final expression for the force required

Not an unreasonable value (~ 0.7 oz)

# EXAMPLE 6.4

A weight of mass $m_1$ hangs from a string that is wrapped around a spool of radius $b$ on a frictionless axle that has a mass of $m_2$ and moment of inertia about its axle of $I = \beta m_2 b^2$. Find an expression for the rotational speed of the spool after the weight has dropped a distance $h$ from rest in terms of $\beta$, $m_1$, $m_2$, $b$, $h$.

| SOLUTION: | ANNOTATION/COMMENT (STEP #): |
|---|---|
| | Before/after problem and changes in speed and position: Use of Conservation of Energy approach is appropriate. |

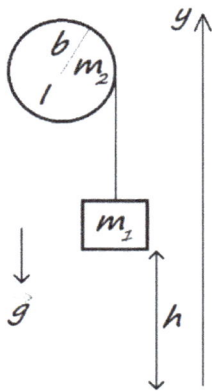

**Draw diagram to help identify all important parts (1)**

$$\begin{pmatrix} 2 \text{ objects} \\ moving \end{pmatrix} \Leftrightarrow \begin{pmatrix} 2 \text{ K.E. terms} \\ per\ side\ of \\ equation \end{pmatrix}$$

$$\begin{pmatrix} 1\ interaction\ by \\ a\ conservative \\ force\ (gravity) \end{pmatrix} \Leftrightarrow \begin{pmatrix} 1\ P.E.\ term \\ per\ side\ of \\ equation \end{pmatrix}$$

$$\begin{pmatrix} No\ non{-} \\ conservative \\ work\ done \end{pmatrix} \Leftrightarrow (no\ W_{nc}\ terms)$$

**Identifying the number of each type of term (1)**

Note: This step is normally not shown but is key to using conservation of energy correctly.

| SOLUTION: | ANNOTATION/COMMENT (STEP #): |
|---|---|

$\omega_i = 0$, $v = b\omega$, $\omega_f = ?$

Let: $y_f = 0$, $y_i = h$

where $y_i = y_f + h$ and $I = \beta m b^2$

**Defining key variables and what is being asked (1)**

$$E_f = E_i + \sum W_{nc}$$
$$K_{m_1,f} + K_{s,f} + U_f = K_{m_1,i} + K_{s,i} + U_i + 0$$
$$K_{m_1,f} + K_{s,f} + 0 = 0 + U_i$$
$$K_{m_1,f} + K_{s,f} = U_i$$

**Filling in the terms in the Conservation of Energy relationship (2)**

<u>Final generic expression</u>

$$\tfrac{1}{2} m_1 v_f^2 + \tfrac{1}{2} I \omega_f^2 = m_1 g y_i$$

**Put in expressions and variables based on the coordinate system being used (3)**

$$\tfrac{1}{2} m_1 (b\omega_f)^2 + \tfrac{1}{2}(\beta m_2 b^2)\omega_f^2 = m_1 gh$$
$$\tfrac{1}{2} m_1 b^2\omega_f^2 + \tfrac{1}{2}\beta m_2 b^2\omega_f^2 = m_1 gh$$
$$\omega_f^2 b^2(m_1 + \beta m_2) = 2m_1 gh$$

**Doing algebra and filling in other relationships (4)**

$$\omega_f^2 = \frac{2m_1 gh}{b^2(m_1 + \beta m_2)}$$

$$\boxed{\omega_f = \frac{1}{b}\left[\frac{2m_1 gh}{(m_1 + \beta m_2)}\right]^{1/2}}$$

<u>Final expression</u>

**Unit analysis (5)**

$$[\omega_f] = \frac{1}{b}\left[\frac{2m_1 gh}{(\beta m_2 + m_1)}\right]^{1/2}$$
$$= \left[\frac{1}{b}\right]\left[\frac{m_1}{(\beta m_2 + m_1)}\right]^{1/2}[2gh]^{1/2}$$
$$= \left(\frac{1}{m}\right)\left[\frac{kg}{(kg + kg)}\right]^{1/2}\left(\left[\frac{m}{s^2}\right]m\right)^{1/2}$$
$$= \left(\frac{1}{\cancel{m}}\right)\left[\frac{\cancel{kg}}{\cancel{kg}}\right]^{1/2}\left(\frac{\cancel{m}}{s}\right)$$
$$[\omega_f] = 1/s \checkmark$$

Note: This step is normally not shown, although it should always be done, at least by inspection. It is a quick way to check if there were any mistakes in the algebra so far.

<u>Units of angular velocity are as expected</u>

## SOLUTION:

for $m_2 \gg m_1$

$$\omega_f = \frac{1}{b}\left[\frac{2m_1 gh}{(\beta m_2 + m_1)}\right]^{1/2}$$

$$= \frac{1}{b}\left[\left(\frac{m_1}{m_2}\right)\frac{2gh}{(\beta + m_1/m_2)}\right]^{1/2}$$

$$\sim \frac{1}{b}\left[\left(\frac{m_1}{m_2}\right)\frac{2gh}{(\beta + 0)}\right]^{1/2}$$

$$\sim \left(\frac{m_1}{m_2}\right)^{1/2}$$

$$\omega_f \sim 0 \checkmark$$

for $m_2 \ll m_1$

$$\omega_f = \frac{1}{b}\left[\frac{2m_1 gh}{(\beta m_2 + m_1)}\right]^{1/2}$$

$$= \frac{1}{b}\left[\frac{2gh}{(\beta(m_2/m_1) + 1)}\right]^{1/2}$$

$$\sim \frac{1}{b}\left[\frac{2gh}{(0 + 1)}\right]^{1/2}$$

$$\sim \frac{1}{b}(2gh)^{1/2}$$

$$\omega_f b \sim (2gh)^{1/2}$$

$$v_f \sim (2gh)^{1/2} \checkmark$$

$$\boxed{\omega_f = \frac{1}{b}\left[\frac{2m_1 gh}{(m_1 + \beta m_2)}\right]^{1/2}}$$

## ANNOTATION/COMMENT (STEP #):

**Check expectations and for physicalness of final expression (4)**

Evaluating limiting behavior for block having negligible mass. Easy to do if expression is rewritten in terms of $m_1/m_2$, which will be about zero.

Note: This step is not normally shown but should be done by inspection to check if the final expression is likely correct.

Would expect hardly any motion for negligible mass of block.

Evaluating limiting behavior for spool having negligible mass. Easy to do if the expression is rewritten in terms of $m_2/m_1$, which will be about zero.

For the spool having negligible mass, the block would accelerate as if in free-fall

Final expression is correct

# EXAMPLE 6.5

A ball is rolling down a ramp inclined at an angle of 25°. Its mass is 450 g, and its radius is 6.5 cm. If it rolls a distance of 20 cm, what is its final angular speed if it was released from rest? What is its angular acceleration as it rolls?

| SOLUTION: | ANNOTATION/COMMENT (STEP #): |
|---|---|
| | Initial/final states and changes in speeds and positions: Use of Conservation of Energy is appropriate. |

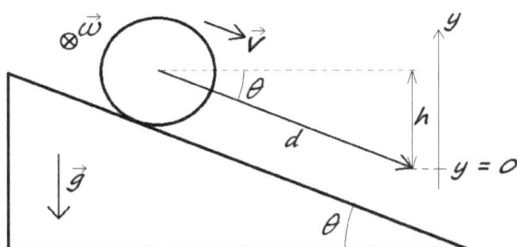

**Diagram to identify key components of system (1)**

$m = 0.45$ kg, $b = 6.5$ cm,

$\theta = 25°$, shape: solid sphere,

$w_i = 0$, $d = 20$ cm

$y_f = 0$, $y_i = h$

**Data for the system components (1)**

$w_f = ?$, $\alpha = ?$

Values to be found

$$E_f = E_i + \sum W_{nc}$$

**Principle to be used (1)**

## SOLUTION:

$$\begin{pmatrix} 1 \text{ object moving} \\ \text{in translation} \\ 1 \text{ object moving} \\ \text{in rotation} \end{pmatrix} \Leftrightarrow \begin{pmatrix} 2 \text{ K.E. terms} \\ \text{per side of} \\ \text{equation} \end{pmatrix}$$

$$\begin{pmatrix} 1 \text{ interaction by} \\ \text{a conservative} \\ \text{force (gravity)} \end{pmatrix} \Leftrightarrow \begin{pmatrix} 1 \text{ P.E. term} \\ \text{per side of} \\ \text{equation} \end{pmatrix}$$

$$\begin{pmatrix} \text{No non-} \\ \text{conservative} \\ \text{work done} \end{pmatrix} \Leftrightarrow (\text{no } W_{nc} \text{ terms})$$

$$E_f = E_i + \sum W_{nc}$$
$$K_{T,f} + K_{R,f} + U_f = K_{T,i} + K_{R,i} + U_i + 0$$
$$\tfrac{1}{2}mv_f^2 + \tfrac{1}{2}I\omega_f^2 + 0 = 0 + 0 + mgh$$
$$\tfrac{1}{2}mv_f^2 + \tfrac{1}{2}I\omega_f^2 = mgh$$

$$v = b\omega, \quad h = d\sin\theta, \quad I_{sph} = \tfrac{2}{5}MR^2$$

$$\tfrac{1}{2}mv_f^2 + \tfrac{1}{2}I\omega_f^2 = mgh$$
$$mv_f^2 + I\omega_f^2 = 2mgh$$
$$m(b\omega_f)^2 + (\tfrac{2}{5}mb^2)\omega_f^2 = 2mgd\sin\theta$$
$$mb^2\omega_f^2 + \tfrac{2}{5}mb^2\omega_f^2 = 2mgd\sin\theta$$
$$(1 + \tfrac{2}{5})mb^2\omega_f^2 = 2mgd\sin\theta$$
$$\tfrac{7}{5}b^2\omega_f^2 = 2gd\sin\theta$$

$$\omega_f^2 = \left(\frac{10}{7}\right)\frac{gd\sin\theta}{b^2}$$

$$\boxed{\omega_f = \left(\left(\frac{10}{7}\right)\frac{gd\sin\theta}{b^2}\right)^{1/2}}$$

## ANNOTATION/COMMENT (STEP #):

**Counting the number of terms for each type of term (1)**

Note: This step is seldom ever shown but is still always done, even if only in one's head.

**Filling in the terms in the general expression (2)**

There are both kinetic energy terms for translation and rotation. These will be connected by $v = b\omega$.

**Extra expressions needed (3)**

**Doing algebra (4)**

Final expression for the angular speed

## SOLUTION:

$$[\omega_f] = \left[\left(\left(\frac{10}{7}\right)\frac{gd\,\sin\theta}{b^2}\right)^{1/2}\right]$$

$$= \left(\left(\left[\frac{10}{7}\right]\right)\frac{[g][d]}{[b^2]}[\sin\theta]\right)^{1/2}$$

$$= \left((1)\frac{(m/s^2)m}{m^2}(1)\right)^{1/2}$$

$$= \left(\frac{m^2}{s^2m^2}\right)^{1/2} = \left(\frac{\cancel{m^2}}{s^2\,\cancel{m^2}}\right)^{1/2}$$

$$[\omega_f] = \frac{1}{s} \checkmark$$

$$\omega_f = \left(\left(\frac{10}{7}\right)\frac{gd\,\sin\theta}{b^2}\right)^{1/2}$$

$$\omega_f \sim \sqrt{d} \checkmark$$
$$\omega_f \sim 1/b \checkmark$$
$$\omega_f \sim \sqrt{\sin\theta} \checkmark$$

$$\omega_f = \left(\left(\frac{10}{7}\right)\frac{gd\,\sin\theta}{b^2}\right)^{1/2}$$

$$v_f = b\omega_f = \left(\left(\frac{10}{7}\right)gd\,\sin\theta\right)^{1/2}$$

## ANNOTATION/COMMENT (STEP #):

### Checking units (5)

Note: This is seldom shown, but even if only done "by inspection," it can be a powerful and simple way to check if any errors have occurred or not.

Units for final angular speed are 1/s as expected.

Note: The units for the final angular velocity will be read as "radians/sec."

### Checking expectations (5)

Note: This step should always be done even if not written out to check for any errors.

The speed is larger for larger $d$, smaller $b$, and larger $\sin\theta$, which would be expected.

The reason for this being inversely proportional to $b$ is that the larger $b$ is, the greater the rotational inertia.

Final expression is rewritten in terms of the more familiar translational speed.

**SOLUTION:**                                    **ANNOTATION/COMMENT (STEP #):**

$$\omega_f = \left(\left(\frac{10}{7}\right)\frac{gd\sin\theta}{b^2}\right)^{1/2}$$

**Evaluating expression numerically (6)**

$$= \left(\frac{10}{7}\right)^{1/2}$$

$$\cdot\left(\left(9.80\frac{m}{s^2}\right)\frac{(0.20\ m)\sin 25°}{(0.065m)^2}\right)^{1/2}$$

$$= 16.7\left(\frac{\cancel{m}}{s^2}\frac{\cancel{m}}{\cancel{m^2}}\right)^{1/2}$$

$$\boxed{\omega_f = 17\frac{1}{s} = 17\frac{rad}{s}}$$

Expression for final angular speed.

$$\alpha = ?$$

**Using kinematics relationship to find acceleration (3)**

$$2\alpha(\theta_f - \theta_i) = \left(\omega_f^2 - \omega_i^2\right)$$

$$d = b(\theta_f - \theta_i)$$

$$2\alpha(\theta_f - \theta_i) = \left(\omega_f^2 - \omega_i^2\right)$$

**Doing algebra (4)**

$$2\alpha\left(\frac{d}{b}\right) = \omega_f^2 - 0$$

$$\alpha = \frac{b\omega_f^2}{2d}$$

Final expression for acceleration.

**Evaluating final expression numerically (6)**

$$\alpha = \frac{b\omega_f^2}{2d}$$

$$= \frac{(0.065\ m)(16.7\ rad/s)^2}{2(0.20\ m)}$$

Unit analysis done during evaluation of final value.

$$\boxed{\alpha = 45\ rad/s^2}$$

Expression for acceleration of sphere. This is the same numerical value as for example 6.2 (to within a rounding error). The lack of a negative sign is due to this example being solved using the Conservation of Energy approach which only includes the magnitudes of the velocities.

# EXAMPLE 6.6

A lump of clay of mass of 35 g moving in the plane perpendicular to the axis of a rotating solid sphere hits and sticks to the surface of the sphere at a distance of 2/5 of its radius from the axis of rotation. If the sphere has a mass of 1.3 kg, a radius of 25 cm, and is rotating at a speed of 1.2 rad/s, how fast must the lump of clay be going to bring the sphere to rest?

## SOLUTION:

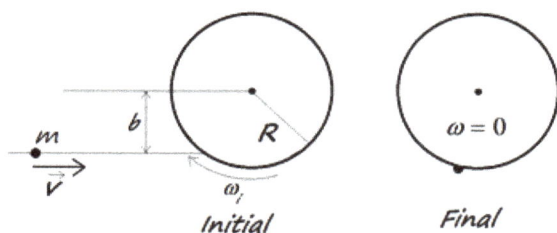

*Initial*          *Final*

$m_{clay} = m = 0.035$ kg

$m_{sphere} = M = 1.3$ kg

$b = (2/5)R$ , $R = 0.25$ m

$\omega_i = -1.2$ rad/s,   $v = ?$

$I_{sph} = \frac{2}{5}MR^2$

$\left(\sum \vec{L}\right)_i = \left(\sum \vec{L}\right)_f$

$\vec{L}_m + \vec{L}_{sph} = \vec{L}_f$

$L_m + L_{sph} = L_f$

$\vec{L}_m = \vec{r} \times \vec{p}$

## ANNOTATION/COMMENT (STEP #):

Initial/Final, velocities and rotating bodies interacting: Use of Conservation of Angular Momentum is appropriate.

**Drawing a diagram to identify the parts of the system. (1)**

For the sphere to come to rest, the lump of clay must be coming from a direction that would counter the rotation of the sphere. This has been incorporated into the setup.

**Listing the data values and value to be found. (1)**

Note: The proper sign of the kinematic parameters (e.g., angular velocities) are important for momentum and angular momentum problems.

**Expression for moment of inertia (1)**

**Applying the physics principle to the problem (2)**

**Scalar version of vector equation (3)**

Expression for angular momentum of lump of clay.

**SOLUTION:**

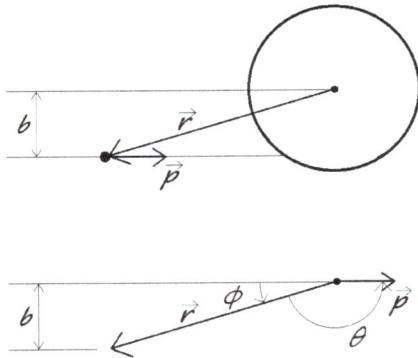

$$\left|\vec{L}_m\right| = \left|\vec{r} \times \vec{p}\right|$$
$$L_m = rp\sin\theta$$
$$= (mv)(r\sin(180° - \phi))$$
$$= (mv)(r\sin(+\phi))$$
$$= mvb$$

$$L_m \rightarrow ccw \text{ rotation, } \therefore L_m > 0$$

$$L_m = +mvb$$

$$L_m + L_{sph} = L_f$$
$$+mvb + I_{sph}\omega_i = I_{tot}\omega_f$$
$$mvb + I_{sph}\omega_i = 0$$
$$mvb = -I_{sph}\omega_i$$
$$v = -\left(\frac{I_{sph}\omega_i}{mb}\right)$$

$$v = -\frac{\left(\frac{2}{5}MR^2\right)\omega_i}{m\left(\frac{2}{5}R\right)} = -\frac{\cancel{\frac{2}{5}}MR^{\cancel{2}}\omega_i}{m\left(\cancel{\frac{2}{5}}R\right)}$$

$$\boxed{v = -\left(\frac{M}{m}\right)R\omega_i}$$

**ANNOTATION/COMMENT (STEP #):**

Diagrams to help in evaluating the magnitude of the cross-product.

Evaluating the cross-product for $L_m$.

Using the relation that $(\theta + \varphi) = 180°$ and that for this situation, $\sin(\theta) = \sin(\varphi)$

With trigonometry, have $b = r\sin(\varphi)$.

Checking sign of $L_m$

Note: For translational and rotational momentum problems, always check the signs for the expressions, as the signs denote direction.

**Doing algebra (4)**

The negative sign captures that the clay must move to oppose the spinning of the sphere.

Substituting the expression for moment of inertia of sphere and for distance $b$.

Final expression for the speed of lump of clay.

| SOLUTION: | ANNOTATION/COMMENT (STEP #): |
|---|---|

**Checking units (4)**

$$[v] = \left[ \left( \frac{M}{m} \right) R\omega_i \right]$$

Note: Checking units should always be done, even if only by inspection, to catch errors before moving on in the problem.

$$= \left[ \frac{M}{m} \right] [R][\omega_i] = \left( \frac{kg}{kg} \right) (m) \left( \frac{1}{s} \right)$$

$$[v] = \frac{m}{s} \checkmark$$

Correct units for angular velocity

**Check predicted behavior (5)**

$v \sim \omega_i \checkmark$

A greater speed is required to stop the sphere, the faster it is going.

$v \sim R \checkmark$

A greater speed is required for a larger sphere due to its greater inertia.

$v \sim (M/m) \checkmark$

A greater speed is required if the inertia of the sphere is large or if the clay is small.

Note: Checking expectations can expose any error that may have been made.

**Evaluating final expression (6)**

$$v = -\frac{MR\omega_i}{m}$$

The signs for the initial angular velocity of the sphere and in the expression for the speed of the clay are canceled.

$$= -\frac{(1.3 \text{ kg})(0.25 \text{ m}) \left( -1.2 \frac{rad}{s} \right)}{0.035 \text{ kg}}$$

$$\boxed{v = +11 \frac{m}{s}}$$

Final answer for speed of the clay.

The speed may seem high, but the inertia of the clay is small, so that great speed would be necessary.

If the sign of the angular velocity of the sphere had been ignored, the velocity of the clay would have been negative, which is not very physical nor consistent with the setup of the problem.

# EXAMPLE 6.7

A solid cone and a disk of equal mass and outer diameter are free to rotate separately about their axes of symmetry on the same axle. The cone is then allowed to drop along the axle, and after a moment, they rotate as one. If the disk is initially rotating at a speed of 1.5 rad/sec and the final speed of the combined object is rotating at a speed of 2.5 rad/sec in the opposite direction, what was the initial rotational speed and direction of rotation of the cone?

**SOLUTION:**

**ANNOTATION/COMMENT (STEP #):**

Initial/Final problem in terms of inertia, velocities, and/or momenta. Use of Conservation of Angular Momentum is appropriate.

**Making a diagram to identify the parts of the system (1)**

Note: Because momentum is a vector quantity, it is important to be explicit with the directions of velocities, not only the magnitudes.

To illustrate how to interpret the signs for velocities, the direction for the unknown velocity, $\omega_c$, was defined to be in the positive direction. Because one the objects in the final state is going in the opposite direction, it would be reasonable to expect $\omega_c$ to be less than zero (in the opposite direction than shown). This will be what is found in the final answer.

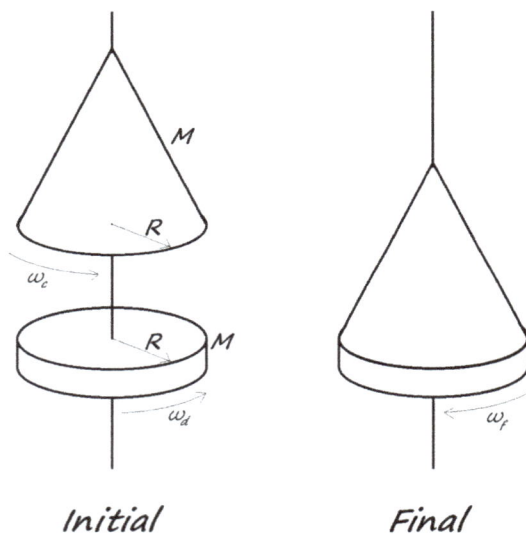

Initial                    Final

**Listing knowns and unknowns (1)**

$\omega_d = +1.5 \text{ rad}/\text{sec}$,
$\omega_f = -2.5 \text{ rad}/\text{sec}$,
$M = m_c = m_d$,
$R = r_d = r_c$

$\omega_c = ?$

**SOLUTION:**

**ANNOTATION/COMMENT (STEP #):**

**Applying principle of conservation of Angular Momentum (2)**

$$\vec{L}_i = \vec{L}_f$$

$$\left(\sum \vec{L}\right)_i = \left(\sum \vec{L}\right)_f$$

Here two objects become one (i.e., a perfectly inelastic collision)

$$\vec{L}_c + \vec{L}_d = \vec{L}_f$$

**Effectively a 1D vector problem, the vector signs are dropped to make it a scalar equation. (3)**

$$L_c + L_d = L_f$$

**Doing algebra (4)**

$$I_c \omega_c + I_d \omega_d = I_f \omega_f$$

$$I_c \omega_c = I_f \omega_f - I_d \omega_d$$

Final expression for angular velocity

$$\boxed{\omega_c = \frac{I_f \omega_f - I_d \omega_d}{I_c}}$$

**Expression for moments of inertia for the two separate objects. (4)**

$$I_d = \tfrac{1}{2} MR^2$$

$$I_c = \tfrac{3}{10} MR^2$$

**Using sum rule for total moment of inertia of final object (4)**

$$I_f = I_c + I_d$$

$$= \left(\tfrac{1}{2} MR^2\right) + \left(\tfrac{3}{10} MR^2\right)$$

$$= \left(\tfrac{1}{2} + \tfrac{3}{10}\right) MR^2 = \tfrac{8}{10} MR^2$$

$$= \tfrac{4}{5} MR^2$$

**Further algebra to simplify the expression for the angular velocity (4)**

$$\omega_c = \frac{I_f \omega_f - I_d \omega_d}{I_c}$$

$$= \frac{\left(\tfrac{4}{5} MR^2\right)\omega_f - \left(\tfrac{1}{2} MR^2\right)\omega_d}{\left(\tfrac{4}{5} MR^2\right)}$$

Eliminating the common factors

$$= \frac{\left(\tfrac{4}{5}\omega_f - \tfrac{1}{2}\omega_d\right)\cancel{MR^2}}{\tfrac{3}{10}\cancel{MR^2}}$$

$$= \left(\frac{10}{3}\right)\left(\frac{4}{5}\omega_f - \frac{1}{2}\omega_d\right)$$

$$\boxed{\omega_c = \frac{8}{3}\omega_f - \frac{5}{3}\omega_d}$$

Final expression for angular velocity

**SOLUTION:**

$$[\omega_c] = \left| \frac{8}{3}\omega_f - \frac{5}{3}\omega_d \right|$$

$$= \left| \frac{8}{3} \right| [\omega_f] - \left| \frac{5}{3} \right| [\omega_d]$$

$$= (1)(1/s) + (1)(1/s)$$

$$[\omega_c] = 1/s \checkmark$$

$$\omega_c = \frac{8}{3}\omega_f - \frac{5}{3}\omega_d$$

$$\omega_c \sim \omega_f - \omega_d \checkmark$$

$$\omega_c = \frac{8}{3}\omega_f - \frac{5}{3}\omega_d$$

$$= \frac{8}{3}\left(-2.5\frac{rad}{s}\right) - \frac{5}{3}\left(+1.5\frac{rad}{s}\right)$$

$$\boxed{\omega_c = -9.2\frac{rad}{s}}$$

**ANNOTATION/COMMENT (STEP #):**

**Unit analysis (5)**

Note: This is an example of an expression for which doing unit analysis is easy to do by inspection.

Units for angular velocity are 1/s as expected.

**Check expectations (5)**

Because the direction of rotation of the combined object is opposite of one of the initial two objects, it makes sense that the final expression would have the functional form of a difference of speeds.

Note: Checking predictions of final expressions is good for catching errors that may have occurred and for checking one's conceptual understanding.

**Putting in data values (6)**

Final numerical answer for the rotational velocity.

The negative value for the velocity of the cone tells us that the original guess for the direction of its velocity was in the opposite direction of what it must be.

The magnitude of the cone's speed is larger, which also makes sense, as it would have to have quite a bit of (rotational) momentum to cause the final combined object to move in the opposite direction of the disk.

# Appendix: Unit Analysis and Unit Conversion

## Introduction

Even though the problem-solving strategies in this book emphasize how to solve problems symbolically, ultimately, the test of any model or theory is done by taking measurements and comparing them to the predictions of the model or theory. As such, understanding units and how to convert them from one type to another is mandatory.[1]

The first of these skills is called "unit analysis," and the second is "unit conversion." Each of these can be defined as follows:

Unit Analysis: The examination of the units in an expression for a quantity (e.g., length, time, mass, etc.) to determine if it is of the form expected: for example, for an expression for the speed of an object, the units would be expected to be of the form length per time.

Unit Conversion: The transformation of an expression in one set of units into another with a new set of units (e.g., miles per hour into meters per second).

This appendix is intended as a brief review in support of the other chapters that focus on solving physics problems using different physics principles. These topics are not difficult but worth the review.

### Skills and Knowledge Needed to Perform Unit Analysis and Unit Conversions

◊ *Algebra skills:* The ability to combine and rearrange an equation using only symbols to represent values.

◊ *Critical thinking*: Knowledge of the unit(s) expected for a physical quantity.

## TARGET AND GOALS

**In this appendix, you will learn the following:**

✔ **How to analyze an expression to check if its units are correct.**

✔ **How to convert the units of an expression into another equivalent set of units.**

## The Nature of Unit Analysis

Physical quantities, irrespective of the system of units used, will have only certain combinations for each type of quantity. For instance, length can be measured in meters or miles, and speed can be measured in meters per second or miles per hour. A list of the fundamental units used in mechanics would consist of length, mass, and time. For convenience, we can represent each of these with the following variable names L, M, and T, respectively. Other common types of quantities consisting of combinations of these units would be speed ($L/T$), acceleration ($L/T^2$), force ($M\,L/T^2$), and energy ($M\,L^2/T^2$).

The usefulness of unit analysis is in how it can check and confirm if an expression has the expected units, and if not, how it then will indicate that an error has occurred in the algebra. This analysis can be performed on both numerical and symbolic expressions. If the expression has the expected units, then the derivation so far is likely (but not guaranteed) to be correct. If the units are not those expected, then a mistake has occurred, and further work is not worth doing until the mistake is corrected. Doing such

---

1   *"I often say that when you can measure what you are speaking about, and express it in numbers, you know something about it; but when you cannot measure it, when you cannot express it in numbers, your knowledge is of a meagre and unsatisfactory kind."* Lord Kelvin 1824–1907, Popular Lectures and Addresses vol. 1 (1889) "Electrical Units of Measurement," delivered May 3, 1883.

checks can save much time when problem-solving and provide confidence that the intermediate steps and the final expression are likely correct.

Doing unit analysis has some special notation and rules. Square brackets around an expression ([…]) are read as "the units of. …" For example, the units of speed, represented by the variable $v$, would look like this: $[v] = [\text{speed}] = L/T$ or just $[v] = L/T$.

Note that numerical values have no unit associated with them; in other words, they are "unitless." A shorthand way of designating this would be $[\#] = 1$, where # is any purely numerical value. This is also true for measures of angles using radians and the arguments and values for many basic mathematical expressions such as trigonometric and logarithmic functions.

## The Steps for Performing Unit Analysis

For products of units: The variable types can be treated much like algebraic variables with regard to canceling.

For example, using the definition of a Joule as $1\ \text{kg} \cdot (\text{m/s})^2$, the units of Joules divided by meters can be simplified as the following to show that it has the units of force ($M \cdot L/T^2$):

$$\left[\frac{J}{m}\right] = [J]\frac{1}{[m]} = \left[\text{kg} \cdot \frac{m^2}{s^2}\right]\frac{1}{[m]} = \frac{[\text{kg}][m^2]}{[s^2][m]} = \frac{[\text{kg}][m]^2}{[s]^2[m]} = \frac{ML^2}{T^2L} = \frac{ML\!\!\!/}{T\!\!\!/} = \frac{M \cdot L}{T^2}$$

For addition or subtraction of units: The terms that make up the sum must have the same units; for example, a speed cannot be added to a distance. However, if all the terms of a sum are the same unit, then the units of that sum can be represented by the one unit symbol. This way, every expression can be reduced to being expressed as a product of units:

What are the units of the following expression $v = \frac{x_f - x_i}{t}$ ?

$$[v] = \left[\frac{x_f - x_i}{t}\right] = \frac{[x_f - x_i]}{[t]} = \frac{([x_f] - [x_i])}{[t]} = \frac{(L+L)}{T} = \frac{L}{T}$$

Below is a table of the combination of units commonly encountered in a course on mechanics, along with their name if they have one.

**TABLE A.1** Various Physical Quantities and Their Units

| physical quantity | type of unit(s) | SI unit(s) | unit name |
|---|---|---|---|
| scalar value, number | 1 (unity) | 1 | unity |
| mass | M | kg | kilogram = kg |
| length | L | m | meter = m |
| time | T | s | second = s |
| speed | L/T | m/s | |
| acceleration | $L/T^2$ | $m/s^2$ | |
| force | $M(L/T^2)$ | $kg \cdot m/s^2$ | Newton = N |
| energy | $M(L/T)^2 = M(L^2/T^2)$ | $kg \cdot m^2/s^2$ | Joule = J |
| momentum | $M(L/T)$ | $kg \cdot m/s$ | |
| torque | $M(L/T^2)L = M(L^2/T^2)$ | $kg \cdot m^2/s^2 = N \cdot m$ | |
| angular momentum | $M(L/T)L = M(L^2/T)$ | $kg \cdot m^2/s$ | |

A note on notation: The standard for notation for symbols used in physics is to use *italics* for variables (e.g., acceleration due to gravity = $g$ or force = $F$), whereas non-italicized letters are reserved for units (e.g., the unit of mass, a kilogram = kg) or for the letters used to describe a type of unit, such as length = L. Look for this throughout this book and your own textbook.

The following examples show several instances of unit analysis. These might seem to show more steps than necessary, but that is purposeful. The intent is to have no ambiguity. In practice, it is acceptable to show fewer steps. And sometimes, the unit analysis can be done simply by inspection, i.e., "in one's head" so that no written work needs to be shown. Whether the unit analysis is shown on paper or not, it should always be done as a quick way to check if the previous work has any errors.

# Steps for Performing Unit Analysis

<u>Hallmarks</u>: Expression with a complicated set of units that needs to be simplified and evaluated to confirm that they are appropriate for the unit(s) expected for the expression of interest.

## Steps

1. **For each variable or unit in an expression, replace it with the appropriate unit variable (e.g., either "L" for length or "m" for meter).**

   - *Separate each factor* in the expression *and insert unit*
   - *Check if factors with sums have the same units*\*
     - *If yes, then combine* and represent that sum with the single unit symbol because the expression's units are not mixed
     - *If not, then STOP*: the expression is not correct

2. **Cancel units and collect the remaining terms**
   - *If units are not compact or obvious, repeat the prior step* to further simplify

3. **Check if the final combination of units is what was expected\***
   - *If yes, then proceed* with the rest of the problem: the prior work is likely correct
   - *If not, STOP* and check prior work: there is an error

(\**These are the most important steps for unit analysis.*)

# EXAMPLE A.1

What are the units of the expression R = 3.0 km/20 mph?

**SOLUTION:**

**ANNOTATION/COMMENT (STEP #):**

$$[R] = \left[\frac{3.0 \ km}{20 \ mph}\right]$$

$$= \frac{[3.0 \ km]}{[20 \ mph]}$$

Factoring expression (1)

$$= \frac{[3.0] \ [km]}{[20] \ [mph]}$$

$$= \frac{[3.0] \ [km]}{[20] \left[\dfrac{mile}{hour}\right]}$$

$$= \frac{[3.0] \ [km]}{[20] \dfrac{[mile]}{[hour]}}$$

$$= \frac{(1)(L)}{(1)(L/T)}$$

Substituting each variable with the symbol for its appropriate type of unit (1)

$$= \frac{L \cdot T}{L}$$

$$= \frac{\cancel{L} \cdot T}{\cancel{L}}$$

Cancel/collect units (2)

$$\boxed{[R] = T} = units \ of \ time$$

Identify the type of unit (3)

# EXAMPLE A.2

Does the following expression have the units of force? $N = mv^2\ell/r^2$

| SOLUTION: | ANNOTATION/COMMENT (STEP #): |
|---|---|

$$[N] = \left| \frac{mv^2\ell}{r^2} \right|$$

$$= \frac{[m][v]^2[\ell]}{[r]^2}$$

Factoring expression (1)

$$= \frac{(M)(L/T)^2(L)}{L^2}$$

Substituting each variable with the symbol for its appropriate type of unit (1)

$$= \frac{ML^3}{L^2T^2}$$

$$= \frac{ML^{\cancel{3}}}{\cancel{L^2}\,T^2}$$

Cancel/collect units (2)

$$= \frac{ML}{T^2}$$

$$\boxed{[N] = M\frac{L}{T^2}} = \text{units of force. Yes}$$

Identify type of unit (3)

# EXAMPLE A.3

Does the following expression have the correct units for acceleration?
$a = g \left( \sin \theta - m_1 \right) / \left( m_1 - m_2 \right)$

**SOLUTION:**

ANNOTATION/COMMENT (STEP #):

$$[a] = \left[ g \left( \frac{\sin \theta - m_1}{m_1 - m_2} \right) \right]$$

$$= [g] \left( \frac{[\sin \theta - m_1]}{[m_1 - m_2]} \right)$$

Factoring expression (1)

$$= \left[ 9.8 \frac{m}{s^2} \right] \frac{[\sin \theta] + [m_1]}{[m_1] + [m_2]}$$

$$= [9.8] \frac{\left[ m \right]}{[s^2]} \frac{[\sin \theta] + [m_1]}{[m_1] + [m_2]}$$

Substituting each variable with the symbol for its appropriate type of unit (1)

$$= 1 \left( \frac{L}{T^2} \right) \frac{(1 + M)}{(M + M)}$$

Cancel/collect units (2)

$$= 1 \left( \frac{L}{T^2} \right) \frac{(1 + M)}{M}$$

$$[a] = \left( \frac{L}{MT^2} \right) (1 + M)$$

Checking units are the same within each factor (3)

<u>No.</u> There are mixed units in the factor
(1 + M), thus the expression cannot
be correct.

<u>One of the factors has a mix of units</u>

# EXAMPLE A.4

Does the expression $v = \sqrt{2gh}$ have the expected units of speed, where $h = 4.3$ m and $g = 9.8$ m/s²?

| SOLUTION: | ANNOTATION/COMMENT (STEP #): |
|---|---|

$$[v] = \left| \sqrt{2gh} \right|$$

$$= \left| \sqrt{2\left(9.8\frac{m}{s^2}\right)(4.3\ m)} \right|$$

$$= \sqrt{[2][9.8]\frac{[m]}{[s^2]}[4.3\ m]}$$

**Factoring expression (1)**

$$= \sqrt{(1)(1)\left(\frac{L}{T^2}\right)(1)(L)}$$

**Substituting each variable with the symbol for its appropriate type of unit (1)**

$$= \sqrt{\left(\frac{L}{T}\right)^2}$$

**Cancel/collect units (2)**

$$[v] = \left|\left(\frac{L}{T}\right)\right| \quad \text{Speed, yes.}$$

**Identify type of unit (3)**

Version of unit analysis with symbolic expression and SI units used.

$$[v] = \left| \sqrt{2gh} \right|$$

$$= \sqrt{[2][g][h]}$$

**Factoring expression (1)**

$$= \sqrt{(1)\left(\frac{m}{s^2}\right)(m)}$$

**Substituting each variable with the symbol for its appropriate type of unit (1)**

$$= \sqrt{\frac{m^2}{s^2}}$$

**Cancel/collect units (2)**

$$[v] = \frac{m}{s} \quad \text{Speed, yes.}$$

**Identify type of unit (3)**

# EXAMPLE A.5

Does the following expression have the expected units of height? $h = v^2/2mg$

**SOLUTION:**

**ANNOTATION/COMMENT (STEP #):**

$$[h] = \left|\frac{v^2}{2mg}\right|$$

$$= \frac{[v]^2}{[2][m][g]}$$

**Factoring expression (1)**

$$= \frac{[m/s]^2}{[2][kg][9.8]\left|\frac{m}{s^2}\right|}$$

(Recall that italicized text is the variable, the non-italicized text is the unit, and the capital non-italicized text is the type of unit.)

$$= \frac{(L/T)^2}{(1)(M)(1)(L/T^2)}$$

**Substituting each variable with the symbol for its appropriate type of unit (1)**

$$= \frac{L^2 T^2}{MLT^2} = \frac{\cancel{L}^2 \cancel{T}^2}{M\cancel{L}\cancel{T}^2}$$

**Cancel/collect units (2)**

$$\boxed{[h] = \frac{L}{M}}$$

**Identify type of unit (3)**

No. Not the units of height.

Here the unit is not what was expected.

# The Nature of Unit Conversion

Unit conversion is the act of re-expressing a quantity in a new set of units (e.g., converting from miles to kilometers) or taking a combination of units and identifying what overall quantity it is expressing (e.g., $1 \text{ kg(m/s)}^2 = 1 \text{ J}$).

Unit conversion will occur in complicated expressions or in various derived values based on the manipulation of a key relationship developed in the act of problem-solving. It is especially important to do this well as the data used to evaluate final expressions may not be of the same unit system (e.g., a mix of SI and Imperial units). If the expressions are developed within a system of units, then the unit conversion process has been embedded in the expression. For instance, take the example of kinetic energy: $K = \frac{1}{2} mv^2$. If the value for the mass is in kg and the speed is in m/s, then the result, by virtue of the fundamental definition of energy, will be in units of Joules. However, if the units are mixed, say $m = 3.4$ kg and $v = 2.2$ miles per hour, then the final combination of units of the kinetic energy, 16 kg (miles$^2$/hour$^2$), is not in units defined for energy in any unit system and will have to be converted to be of practical use.

## Comments on the Process of Unit Conversion

- Even with the most complete table of conversions, it is not always possible to find a factor that will allow you to do a conversion in one step. This is ok.

- Using the product of multiple factors for doing the conversion can be easier instead of using one seldom used conversion that must be remembered. Using several simple ones that are more easily remembered allows you to convert an expression faster without referring to an external source.

- Doing unit analysis at the end of your conversion is a quick way to check your answer for errors. If the units are not what you expect, then there was likely an error that should be fixed before moving on.

- There is no fixed order to which the factors used in the conversion must be executed. Like with algebra, if each step is done properly, the outcome will be correct whatever the path taken.

# The Steps for Performing Unit Conversion

For simple conversions, a direct substitution of a unit with its equivalent can work. For example, $\ell = 36$ inches is converted to feet by substituting 1 inch for $(^1/_{12})$ foot so that $\ell = 36 ((^1/_{12})(\text{foot})) = 3$ feet.

For more complicated expressions and for ones for which you will want to show a record of the conversion, a better way is the "*multiply by unity*" method. For this, the expression is multiplied by a factor of one but expressed as a ratio of two units: For example, 1.00 mile = 1.61 km can be rewritten as either

$$1 = \left( \frac{1.61 \text{ km}}{1.00 \text{ mile}} \right) \quad \text{or} \quad 1 = \left( \frac{1.00 \text{ mile}}{1.61 \text{ km}} \right)$$

The choice of which to use will depend on the units to be canceled.

For example, to convert a distance of 9.2 miles to kilometers, we would do the following:

$$d = 9.2 \text{ miles} = 9.2 \text{ miles} \left( \frac{1.61 \text{ km}}{1.00 \text{ miles}} \right) = 9.2 \text{ miles} \left( \frac{1.61 \text{ km}}{1.00 \text{ miles}} \right) = 14.8 \text{ km}$$

For the above example, the 9.2 miles was multiplied by 1 = (1.61 km/1.00 mile) so that the miles would cancel.

If the reciprocal had been chosen, the expression would have technically still been correct because mathematically we had only multiplied it by 1, and by unit analysis it would have represented a length ($L^2/L = L$). However, it would have been of little practical use. In other words,

$$d = 9.2 \text{ miles} = 9.2 \text{ miles} \left( \frac{1.00 \text{ miles}}{1.61 \text{ km}} \right) = 5.7 \frac{\text{miles}^2}{\text{km}}.$$

The following examples, like most in this book, are written showing many steps, likely more than most would use. The reason is so that there is no ambiguity as to how the conversion was done. In practice, whether for an assignment or for your own professional work, you will want to record at the minimum an abbreviated form of your unit conversion process. This will allow you to find your own mistakes and show enough work to receive full credit from your grader. For your professional work, you will want to have a record of your work so that you can recreate it for your colleagues and justify your conclusions.

Regarding the use of online conversion tools: These are useful and can provide a single conversion relationship instead of the several used in some of the examples. It is perfectly appropriate to use them in practice and to use the one conversion equation in your conversion calculations. However, much like other problem-solving tasks, it is the practice of working solutions in a thorough manner that will give you the confidence to do unit conversions on more complex sets of units, for example, electromagnetic units. Use the opportunity you have now to learn to do your unit conversions with a minimum of shortcuts, and you will have an easier time later.

### A Cautionary Tale of Improper Unit Conversion

Perhaps the most famous example of a problem associated with unit conversion is the fate of the Mars Climate Orbiter launched on December 11, 1998. Because of miscommunication between the engineers at Lockheed-Martin and NASA, the units used by each for rocket impulse differed. Lockheed-Martin used pound-force·seconds for its calculations and the computer programs created by NASA for operating the thrusters were written, assuming the units had been converted to Newton·seconds. The result was that the probe came too close to the atmosphere of Mars and burned up. One of the following examples works out this conversion.

# Steps for Performing Unit Conversion

<u>Hallmarks</u>: An expression is desired to be transformed from being expressed in one set of units into another set of units.

## Steps

1. **Identify desired units and unit conversions needed**
2. **Transform units**

   - *Multiply original expression by conversion factors* that equal unity and with the factors orientated so that the desired units cancel

   - *Cancel* the intermediate units

3. **Collect factors from numerators and denominators into final form and multiply by numerical factors**

   - *Check the final units* to see if they are as expected*

     - *If yes, proceed* with the rest of the problem. Prior work is likely correct.

     - *If no, STOP*. There is an error in the prior work.

(**This is the most important step in unit conversion.*)

# EXAMPLE A.6

The length of a rope is 5.55 meters. What is its length in feet? What is the length in feet and inches?

**SOLUTION:**

$12'' = 1 \text{ ft}, \ 1'' = 2.54 \text{ cm}, \ 100 \text{ cm} = 1 \text{ m}$ — **List of unit conversions to use (1)**

$\ell = 5.55 \text{ m} \left( \dfrac{100 \text{ cm}}{1 \text{ m}} \right)$ — **Multiply by conversion factors and cancel common factors (2)**

$\cdot \left( \dfrac{1''}{2.54 \text{ cm}} \right) \left( \dfrac{1 \text{ ft}}{12''} \right)$

$= 5.55 \ \cancel{\text{m}} \left( \dfrac{100 \ \cancel{\text{cm}}}{1 \ \cancel{\text{m}}} \right)$ — **Canceling the intermediate units (2)**

$\cdot \left( \dfrac{1''}{2.54 \ \cancel{\text{cm}}} \right) \left( \dfrac{1 \text{ ft}}{12''} \right)$ — (Note use of different types of slashes to more easily keep track of which canceled units are paired.)

$= 5.55 \left( \dfrac{100 \cdot 1 \cdot 1}{1 \cdot 2.54 \cdot 12} \right) \text{ft}$

$\boxed{\ell = 18.2 \text{ ft}}$ — **Evaluation of expression (3)**

$\ell = 18.2 \text{ ft} = 18 \text{ ft} + 0.2 \text{ ft}$ — Identifying remainder part to evaluate.

$12'' = 1 \text{ ft}$ — **List of unit conversions to use (1)**

$R = 0.2 \text{ ft} = 0.2 \text{ ft} \left( \dfrac{12''}{1 \text{ ft}} \right)$ — **Multiply by conversion factors and cancel common factors (2)**

$= 0.2 \ \cancel{\text{ft}} \left( \dfrac{12''}{1 \ \cancel{\text{ft}}} \right) = 0.2 \cdot 12''$

$\boxed{R = 2''}$ — **Evaluation of expression (3)**

$\boxed{\ell = 18 \text{ ft } \ 2''}$ — **Evaluation of final expression (3)**

A car is traveling at 32 miles per hour. What is its speed in meters per second?

**SOLUTION:**                                          **ANNOTATION/COMMENT (STEP #):**

1 mile = 1.61 km

1 km = 1,000 m                                         List of unit conversions to use (1)

1 hr = 60 minutes

1 minute = 60 seconds

$$v = 32 \text{ m.p.h.} = 32 \frac{miles}{hour}$$

$$v = 32 \frac{mi}{hr} \left( \frac{1.61 \text{ km}}{mi} \right) \cdot$$

$$\cdot \left( \frac{1,000 \text{ m}}{1 \text{ km}} \right) \left( \frac{1 \text{ hr}}{60 \text{ min}} \right) \left( \frac{1 \text{ min}}{60 \text{ s}} \right)$$

Multiply by conversion factors (2)

$$v = 32 \frac{\cancel{mi}}{\cancel{hr}} \left( \frac{1.61 \cancel{km}}{\cancel{mi}} \right) \cdot$$

Canceling the intermediate units (2)

$$\cdot \left( \frac{1,000 \text{ m}}{1 \cancel{km}} \right) \left( \frac{1 \cancel{hr}}{60 \cancel{min}} \right) \left( \frac{1 \cancel{min}}{60 \text{ s}} \right)$$

$$v = 32 \left( \frac{1.61 \cdot 1,000}{60 \cdot 60} \right) \frac{m}{s}$$

$$\boxed{v = 14 \frac{m}{s}}$$

Evaluation of final expression (3)

# EXAMPLE A.8

What is the kinetic energy in Joules of a projectile with a mass of 150 grains traveling at 2,200 ft/s?

| SOLUTION: | ANNOTATION/COMMENT (STEP #): |
|---|---|

$1 \text{ J} = 1 \text{ kg·m}^2/\text{s}^2$

$1 \text{ m} = 100 \text{ cm}$

**List of unit conversions to use (1)**

$2.54 \text{ cm} = 1 \text{ in}, 1 \text{ ft} = 12 \text{ in}$

$1 \text{ gr} = 0.065 \text{ g}, 1{,}000 \text{ g} = 1 \text{ kg}$

For exact definitions units, the precision of the value is treated as infinite. E.g., 1 inch is defined to be exactly 2.54 cm.

$$K = \frac{1}{2}mv^2$$
$$= \frac{1}{2}(150 \text{ gr})(2{,}200 \text{ ft/s})^2$$
$$= \frac{1}{2}(150)(2{,}200)^2 (\text{gr})(\text{ft/s})^2$$

$$K = 3.63 \times 10^8 \text{ gr} \cdot \frac{\text{ft}^2}{\text{s}^2}$$

$$K = 3.63 \times 10^8 \text{ gr} \cdot \frac{\text{ft}^2}{\text{s}^2}$$

**Multiply by conversion factors (2)**

$$\cdot \left(\frac{1 \text{ kg}}{1{,}000 \text{ g}}\right)\left(\frac{0.065 \text{ g}}{1 \text{ gr}}\right)$$

$$\cdot \left(\frac{12 \text{ in}}{1 \text{ ft}}\right)^2 \left(\frac{2.54 \text{ cm}}{1 \text{ in}}\right)^2$$

$$\cdot \left(\frac{1 \text{ m}}{100 \text{ cm}}\right)^2$$

Note how values that are squared have been treated by squaring both the prefactor and the unit. A common mistake is to overlook squaring the prefactor along with the unit.

$$= 3.63 \times 10^8 \text{ gr} \cdot \frac{\text{ft}^2}{\text{s}^2}$$

$$\cdot \left(\frac{1 \text{ kg}}{1{,}000 \text{ g}}\right)\left(\frac{0.065 \text{ g}}{1 \text{ gr}}\right)$$

$$\cdot \left(\frac{144 \text{ in}^2}{1 \text{ ft}^2}\right)\left(\frac{6.54 \text{ cm}^2}{1 \text{ in}^2}\right)$$

$$\cdot \left(\frac{1 \text{ m}^2}{10{,}000 \text{ cm}^2}\right)$$

**SOLUTION:**

**ANNOTATION/COMMENT (STEP #):**

$$K = 3.63 \times 10^8 \; \cancel{gr} \cdot \frac{\cancel{ft^2}}{s^2}$$

**Canceling of common units (2)**

$$\cdot \left( \frac{1 \; kg}{1{,}000 \; \cancel{g}} \right) \left( \frac{0.065 \; \cancel{g}}{1 \; \cancel{gr}} \right)$$

$$\cdot \left( \frac{144 \; \cancel{in^2}}{1 \; \cancel{ft^2}} \right) \left( \frac{6.54 \; \cancel{cm^2}}{1 \; \cancel{in^2}} \right)$$

$$\cdot \left( \frac{1 \; m^2}{10{,}000 \; \cancel{cm^2}} \right)$$

$$K = 3.63 \times 10^8$$
$$\cdot \left( \frac{1}{1{,}000} \right) \left( \frac{0.065}{1} \right)$$

$$\cdot \left( \frac{144}{1} \right) \left( \frac{6.54}{1} \right) \left( \frac{1}{10{,}000} \right) kg \frac{m^2}{s^2}$$

$$K = 3.63 \times 10^8$$
$$\cdot \left( \frac{0.065 \cdot 144 \cdot 6.54}{10^3 \cdot 10^4} \right) kg \frac{m^2}{s^2}$$

**Evaluation of final expression (3)**

$$K = 3.63 \times 10^8 \left( \frac{60.4}{10^7} \right) kg \frac{m^2}{s^2}$$
$$K = 2{,}190 \; kg \frac{m^2}{s^2}$$

Further conversion from basic SI units into Joules.

$$K = 2{,}190 \left( kg \frac{m^2}{s^2} \right) \left( 1J \Big/ \left( kg \frac{m^2}{s^2} \right) \right)$$

**Multiply by conversion factors and cancel common factors (2)**

$$K = 2{,}190 \left( \cancel{kg \frac{m^2}{s^2}} \right) \left( 1J \Big/ \left( \cancel{kg \frac{m^2}{s^2}} \right) \right)$$
$$K = 2{,}190 \; J$$
$$\boxed{K = 2.2 \; kJ}$$

**Evaluation of final expression (3)**

| SOLUTION: | ANNOTATION/COMMENT (STEP #): |
|---|---|

Alternate method: This conversion could also have been done by converting each of the provided data values into SI units and then using those converted values to compute the kinetic energy.

As with algebra, the specific order of steps is not important so long as each step is done correctly.

$1\ m = 3.28\ ft$

$v = 2,200\ \dfrac{ft}{s}$

**List of unit conversions to use (1)**

$= 2,200\left(\dfrac{\cancel{ft}}{s}\right)\left(\dfrac{m}{3.28\ \cancel{ft}}\right)$

**Multiply by conversion factors and cancel common factors (2)**

$\boxed{v = \dfrac{2,200}{3.28}\dfrac{m}{s} = 671\ \dfrac{m}{s}}$

**Evaluation of (intermediate) expression (3)**

$1\ gr = 0.065\ g,\ 1\ kg = 1,000\ g$
$m = 150\ gr$

**List of unit conversions to use (1)**

$= 150\ \cancel{gr}\left(\dfrac{0.065\ \cancel{g}}{1\ \cancel{gr}}\right)\left(\dfrac{1\ kg}{1,000\ \cancel{g}}\right)$

**Multiply by conversion factors and cancel common factors (2)**

$m = \dfrac{150\cdot 0.065\cdot 1}{1\cdot 1,000}kg$

**Evaluation of (intermediate) expression (3)**

$\boxed{m = 9.75\times 10^{-3}\ kg}$

$K = \tfrac{1}{2}mv^2$

$= \tfrac{1}{2}\left(9.75\times 10^{-3}kg\right)\left(671\ \dfrac{m}{s}\right)^2$

$= \tfrac{1}{2}\left(9.75\times 10^{-3}\right)(671)^2 kg\left(\dfrac{m^2}{s^2}\right)$

**Multiply by conversion factors and cancel common factors (2)**

$= 2,190\left(\dfrac{kg\ \cancel{m^2}}{\cancel{s^2}}\right)\left(J\Big/kg\dfrac{\cancel{m^2}}{\cancel{s^2}}\right)$

**Evaluation of final expression (3)**

$\boxed{K = 2.2\ kJ}$

# EXAMPLE A.9

What is an impulse, $\Delta p$, of one pound–force·second (1.0 lbf·s) in Newton·seconds (N·s)? (This is the conversion the people at NASA and Lockheed–Martin missed.)

**SOLUTION:**

**ANNOTATION/COMMENT (STEP #):**

$$1 \text{ lbf} = (1 \text{ slug})\left(1\frac{ft}{s^2}\right) = \frac{1 \text{ slug} \cdot ft}{s^2}$$

**List of unit conversions to use (1)**

$1 \text{ slug} = 14.59 \text{ kg}$

$1 \text{ ft} = 12 \text{ in.}, 1 \text{ in} = 2.54 \text{ cm}$

$100 \text{ cm} = 1 \text{ m}$

The definition for one pound-force of force is read as "a force of 1 pound-force is what is required to accelerate a mass of 1 slug at $1 \text{ ft/s}^2$."

$\Delta p = 1 \text{ lbf} \cdot s$

$$= 1 \text{ lbf} \cdot s\left(slug\frac{ft}{s^2}\middle/lbf\right)$$

**Multiply by conversion factors (2)**

$$\cdot\left(\frac{14.6 \text{ kg}}{1 \text{ slug}}\middle|\frac{12 \text{ in}}{1 \text{ ft}}\right)$$

$$\cdot\left(\frac{2.54 \text{ cm}}{1 \text{ in}}\middle|\frac{1 \text{ m}}{100 \text{ cm}}\right)$$

$$= 1 \cancel{\text{lbf}} \cdot s\left(\cancel{slug}\frac{\cancel{ft}}{s^2}\middle/\cancel{lbf}\right)$$

**Canceling of common units (2)**

$$\cdot\left(\frac{14.6 \text{ kg}}{1 \cancel{slug}}\middle|\frac{12 \cancel{in}}{1 \cancel{ft}}\right)$$

$$\cdot\left(\frac{2.54 \cancel{cm}}{1 \cancel{in}}\middle|\frac{1 \text{ m}}{100 \cancel{cm}}\right)$$

$$= \frac{1(14.6)(2.54)(12)}{100} \cdot \frac{s \cdot m \cdot kg}{s^2}$$

$$= 4.45 \text{ kg}\frac{m}{s^2} \cdot s$$

$$= 4.45\left(\cancel{kg}\frac{\cancel{m}}{\cancel{s^2}}\middle|1N\middle/\cancel{kg}\frac{\cancel{m}}{\cancel{s^2}}\right) \cdot s$$

**Multiply by conversion factors and cancel common factors (2)**

$$\boxed{\Delta p = 4.45 \text{ N} \cdot s}$$

**Evaluation of final expression (3)**

www.ingramcontent.com/pod-product-compliance
Lightning Source LLC
Chambersburg PA
CBHW042031220326
41598CB00073BA/7451